# Dengue Virus Disease

# Dengue Virus Disease

## From Origin to Outbreak

*Edited by*

**Adnan I. Qureshi**
Executive Director, Zeenat Qureshi Stroke Institute,
St. Cloud, MN, United States

Professor, Department of Neurology at University of Missouri,
Columbia, United States

**Omar Saeed**
Clinical Research Fellow, Zeenat Qureshi Stroke Institute,
St. Cloud, MN, United States

Resident Physician, University of Tennessee Health Science Center in
Memphis, Tennessee, United States

ELSEVIER

ACADEMIC PRESS
An imprint of Elsevier

Academic Press is an imprint of Elsevier
125 London Wall, London EC2Y 5AS, United Kingdom
525 B Street, Suite 1650, San Diego, CA 92101, United States
50 Hampshire Street, 5th Floor, Cambridge, MA 02139, United States
The Boulevard, Langford Lane, Kidlington, Oxford OX5 1GB, United Kingdom



**Library of Congress Cataloging-in-Publication Data**
A catalog record for this book is available from the Library of Congress

**British Library Cataloguing-in-Publication Data**
A catalogue record for this book is available from the British Library

ISBN: 978-0-12-818270-3

For information on all Academic Press publications visit our website at https://www.elsevier.com/books-and-journals

*Publisher:* Andre Gerhard Wolff
*Acquisition Editor:* Kattie Washington
*Editorial Project Manager:* Anna Dubnow
*Production Project Manager:* Sreejith Viswanathan
*Cover Designer:* Mark Rogers

Typeset by TNQ Technologies

# Contents

## 5. Mosquito-borne diseases

*Muhammad A. Saleem and Iryna Lobanova*



## 6. Viral genetics and structure

*Iqra Naveed Akhtar*

# Contributors

**Iqra Naveed Akhtar**, Clinical Research Fellow, Department of Neurology, University of Missouri, Columbia, MO, United States; Zeenat Qureshi Stroke Institute, St. Cloud, MN, United States

**Mohammad Ali Arif**, Internal Medicine Shaheed Zulfiqar Ali Bhutto Medical University/Pakistan Institute of Medical Sciences, Islamabad, Punjab, Pakistan; Internal Medicine, Ali Medical Centre, Islamabad, Punjab, Pakistan

**Ahmer Asif**, Zeenat Qureshi Stroke Institute, St. Cloud, MN, United States; Department of Internal Medicine, University of Oklahoma, Oklahoma City, United States

**Sachin M. Bhagavan**, Resident Physician, Department of Neurology, University of Missouri Health Care, Columbia, MO, United States

**Mohammad Rauf A. Chaudhry**, Resident Physician, Texas Tech University Health Sciences Center, El Paso, TX, United States; Zeenat Qureshi Stroke Institute, St. Cloud, MN, United States

**Mohammad F. Ishfaq**, Resident physician, University of Tennessee Health Science center, Memphis, Tennessee, United States; Zeenat Qureshi Stroke Institute, St. Cloud, MN, United States

**Jahanzeb Liaqat**, Lt. Colonel, Pak Emirates, Military Hospital, Rawalpindi, Pakistan

**Iryna Lobanova**, Project Manager, Dengue virus disease project, Zeenat Qureshi Stroke Institute, St. Cloud, MN, United States; University of Missouri, Columbia, MO, United States

**Ngan Nguyen**, Department of Internal Medicine, University of Tennessee Health Science Center, Memphis, Tennessee, United States

**Adnan I. Qureshi**, Executive Director, Zeenat Qureshi Stroke Institutes, St. Cloud, MN, United States; Professor, Department of Neurology at University of Missouri, Columbia, United States

**Ihtesham Qureshi**, Resident Physician, Department of Neurology, Texas Tech University of Health Sciences Center, Paul L. Foster School of Medicine, El Paso, TX, United States

**Mushtaq H. Qureshi**, Texas Tech University Health Sciences Center El Paso, Neurology, Texas Tech University, El Paso, TX, United States; Zeenat Qureshi Stroke Institute, St. Cloud, MN, United States

**Arbaab Qureshi**, Clinical Research Fellow, Department of Neurology, Texas Tech University of Health Sciences, El Paso, TX, United States

**Omar Saeed**, Resident Physician, Department of Neurology, University of Tennessee Health Science Center, Memphis, Tennessee, United States; Zeenat Qureshi Stroke Institute, St. Cloud, MN, United States

**Muhammad A. Saleem**, Resident Physician, Family Medicine, Mercyhealth, Janesville, WI, United States

**Sargun Singh Walia**, Clinical Research Fellow, Zeenat Qureshi Stroke Institute, St. Cloud, MN, United States; Department of Neurology, University of Missouri, Columbia, MO, United States

Chapter 1

# Dengue virus infection

**Adnan I. Qureshi**
*Executive Director, Zeenat Qureshi Stroke Institute, St. Cloud, MN, United States; Professor, Department of Neurology at University of Missouri, Columbia, United States*

## Introduction

The World Health Organization website states "Dengue is a mosquito-borne viral disease that has rapidly spread in all regions of World Health Organization in recent years". In April 2016, World Health Organization issued a conditional recommendation on the use of the vaccine for areas, in which Dengue virus infection is highly endemic, which is defined by population seroprevalence of 70% or higher. Dengue virus infection mainly causes a self-limiting flu-like illness and may remain asymptomatic (Fig. 1.1). A reader may wonder why a book should be dedicated to this disease. There are several reasons: 1. The infection can develop into a potentially severe Dengue virus illness which can be fatal and now a major cause of severe illness among children; 2. The disease is increasing exponentially in prevalence; 3. There is no specific treatment for the disease. The global incident map website provides a detailed list of hundreds of Dengue viral illness outbreaks since 2010. The first and last page of the listing is provided to give a better perspective of large number of incident cases in the world. The global map from new Dengue virus disease cases from May 2019 also provides a regional perspective on the disease (Fig. 1.2).

Health officials in the Philippines in July 2019 declared a national emergency after a record-breaking 106,630 cases of Dengue viral fever were reported since January 2019 [1]. This represented a 85% increase than the number of cases in the same period in 2018. Approximately 500 people have died from the Dengue viral illness in 2019. Military hospitals and clinics were put on alert for a possible surge in Dengue viral illness patients [2].

World Health Organization/Department of Control of Neglected Tropical Diseases released a detailed document "The Global Strategy for dengue prevention and control, 2012–20". (Fig. 1.3) The executive summary of the document states:

"Dengue is a major public-health concern throughout tropical and subtropical regions of the world. It is the most rapidly spreading mosquito-

**Dengue Virus Disease. https://doi.org/10.1016/B978-0-12-818270-3.00001-1**

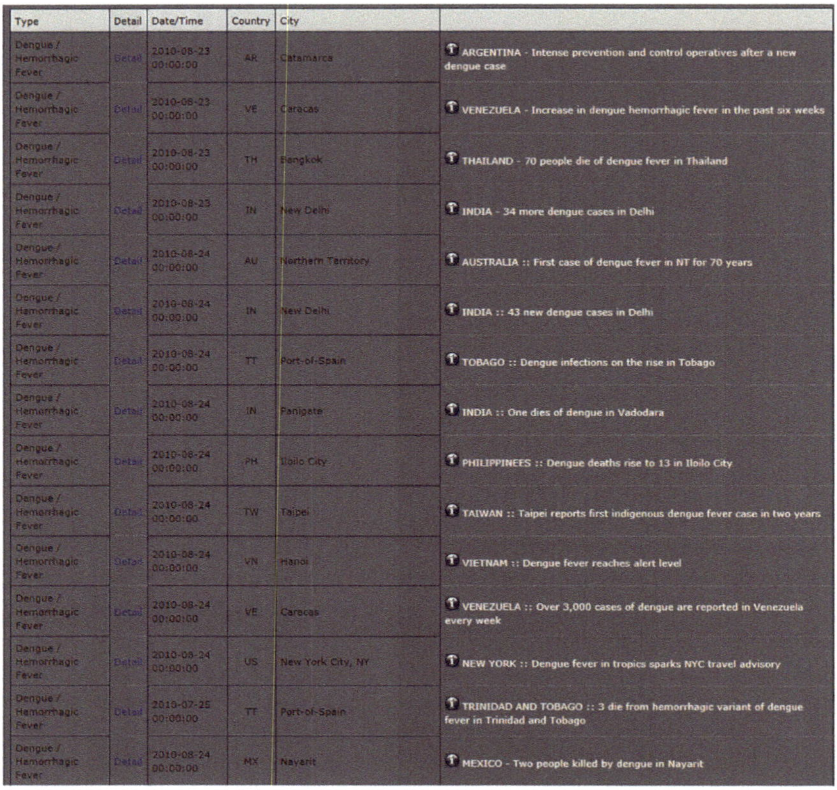

| Type | Detail | Date/Time | Country | City | |
|---|---|---|---|---|---|
| Dengue / Hemorrhagic Fever | Detail | 2010-08-23 00:00:00 | AR | Catamarca | ARGENTINA - Intense prevention and control operatives after a new dengue case |
| Dengue / Hemorrhagic Fever | Detail | 2010-08-23 00:00:00 | VE | Caracas | VENEZUELA - Increase in dengue hemorrhagic fever in the past six weeks |
| Dengue / Hemorrhagic Fever | Detail | 2010-08-23 00:00:00 | TH | Bangkok | THAILAND - 70 people die of dengue fever in Thailand |
| Dengue / Hemorrhagic Fever | Detail | 2010-08-23 00:00:00 | IN | New Delhi | INDIA - 34 more dengue cases in Delhi |
| Dengue / Hemorrhagic Fever | Detail | 2010-08-24 00:00:00 | AU | Northern Territory | AUSTRALIA :: First case of dengue fever in NT for 70 years |
| Dengue / Hemorrhagic Fever | Detail | 2010-08-24 00:00:00 | IN | New Delhi | INDIA :: 43 new dengue cases in Delhi |
| Dengue / Hemorrhagic Fever | Detail | 2010-08-24 00:00:00 | TT | Port-of-Spain | TOBAGO :: Dengue infections on the rise in Tobago |
| Dengue / Hemorrhagic Fever | Detail | 2010-08-24 00:00:00 | IN | Panigate | INDIA :: One dies of dengue in Vadodara |
| Dengue / Hemorrhagic Fever | Detail | 2010-08-24 00:00:00 | PH | Iloilo City | PHILIPPINEES :: Dengue deaths rise to 13 in Iloilo City |
| Dengue / Hemorrhagic Fever | Detail | 2010-08-24 00:00:00 | TW | Taipei | TAIWAN :: Taipei reports first indigenous dengue fever case in two years |
| Dengue / Hemorrhagic Fever | Detail | 2010-08-24 00:00:00 | VN | Hanoi | VIETNAM :: Dengue fever reaches alert level |
| Dengue / Hemorrhagic Fever | Detail | 2010-08-24 00:00:00 | VE | Caracas | VENEZUELA :: Over 3,000 cases of dengue are reported in Venezuela every week |
| Dengue / Hemorrhagic Fever | Detail | 2010-08-24 00:00:00 | US | New York City, NY | NEW YORK :: Dengue fever in tropics sparks NYC travel advisory |
| Dengue / Hemorrhagic Fever | Detail | 2010-07-25 00:00:00 | TT | Port-of-Spain | TRINIDAD AND TOBAGO :: 3 die from hemorrhagic variant of dengue fever in Trinidad and Tobago |
| Dengue / Hemorrhagic Fever | Detail | 2010-08-24 00:00:00 | MX | Nayarit | MEXICO - Two people killed by dengue in Nayarit |

FIGURE 1.1 List of country with Dengue hemorrhagic fever [9].

FIGURE 1.2 Map of Dengue virus disease outbreak around the world [8].

| | | | | |
|---|---|---|---|---|
| 2019-05-20 22:04:00 | Detail | Thailand | Bangkok | THAILAND - Thailand Records More Than 20000 Cases Of Dengue Fever In Almost 5 Months |
| 2019-05-20 21:57:00 | Detail | French Polynesia | Papeete | FRENCH POLYNESIA - Dengue Type 2 Spreads In French Polynesia |
| 2019-05-19 22:47:00 | Detail | Honduras | Tegucigalpa | HONDURAS - 24 Dengue Deaths Are Counted In The Country This Year |
| 2019-05-19 22:00:00 | Detail | Nepal | Dharan | NEPAL - Nine New Cases Of Dengue Reported In Dharan |
| 2019-05-18 23:09:00 | Detail | Argentina | Jujuy | ARGENTINA - They Amount To 283 Positive Cases Of Dengue |
| 2019-05-18 23:05:00 | Detail | Brazil | State Of Rio Grande Do Sul | BRAZIL - RS Has 456 Records Of Dengue Contracted In The State |
| 2019-05-18 22:40:00 | Detail | Argentina | Buenos Aires | ARGENTINA - There Are Already 131 Cases Of Dengue And Two More Have DEN-4 |
| 2019-05-18 22:35:00 | Detail | Chile | Easter Island | CHILE - Health Warning In Rapa Nui After Detection Of 17 Cases Of Dengue |
| 2019-05-18 00:05:00 | Detail | United States | Shelby County, TN | TENNESSEE - Equine Infectious Anemia Reported In Tennessee |
| 2019-05-17 23:18:00 | Detail | Brazil | Cianorte | BRAZIL - Dengue Cases Reach 449 |
| 2019-05-17 22:32:00 | Detail | Brazil | State Of Santa Catarina | BRAZIL - Number Of Dengue Cases In SC Reaches 573 In 2019 |
| 2019-05-17 22:26:00 | Detail | India | Kozhikode | INDIA - Four Dengue Cases In May |
| 2019-05-17 22:22:00 | Detail | Brazil | Umuarama | BRAZIL - Up To 84 Confirmed Cases Of Dengue In Umuarama |
| 2019-05-17 22:20:00 | Detail | Colombia | Bucaramanga | COLOMBIA - In Study - 3 Possible Deaths Due To Dengue |
| 2019-05-17 22:16:00 | Detail | Dominican Republic | Santo Domingo | DOMINICAN REPUBLIC - They Register One Thousand 807 Cases Of Dengue In The Dominican Republic |
| 2019-05-17 22:14:00 | Detail | Brazil | Sao Jose do Rio Preto | BRAZIL - Rio Preto Has More Than 14 Thousand Positive Cases Of Dengue In The Year |
| 2019-05-17 22:11:00 | Detail | India | Krishna | INDIA - 30 Dengue Hot Spots Identified In Krishna District |
| 2019-05-17 22:09:00 | Detail | India | Kozhikode | INDIA - Four Confirmed Dengue Cases Reported |
| 2019-05-17 22:06:00 | Detail | India | Karnataka | INDIA - 383 Dengue Cases Recorded In Bengaluru In 5 Months |
| 2019-05-17 22:02:00 | Detail | Singapore | Pulau Ujong | SINGAPORE - Number Of Dengue Cases In 2019 Surpasses Last year Total Count |
| 2019-05-16 23:32:00 | Detail | Bolivia | La Paz | BOLIVIA - Confirmed 495 Cases Of Dengue In La Paz |
| 2019-05-16 23:21:00 | Detail | Argentina | Chaco Province | ARGENTINA - 60 Cases Of Dengue Confirmed And 140 Still Under Study |
| 2019-05-16 23:16:00 | Detail | Brazil | Cascavel | BRAZIL - Bulletin Points To 579 Positive Cases Of Dengue In Cascavel |
| 2019-05-16 23:02:00 | Detail | Brazil | State Of Rio Grande Do Sul | BRAZIL - Cases Of Dengue Contracted Within The RS Reach 456 In 2019 |
| 2019-05-16 22:51:00 | Detail | Nepal | Dharan | NEPAL - Three Children Infected With Dengue In Dharan |
| 2019-05-16 22:23:00 | Detail | Brazil | Capivari | BRAZIL - Capivari Registers 111 Positive Cases Of Dengue Fever |
| 2019-05-16 21:44:00 | Detail | Tanzania | Dodoma | TANZANIA - Tanzania Fights Outbreak As Dengue Hits Three Regions |
| 2019-05-15 23:12:00 | Detail | Brazil | Cabreuva | BRAZIL - Cabreuva Had 9 Confirmed Cases Of Dengue This Year |
| 2019-05-15 23:07:00 | Detail | Brazil | Presidente Prudente | BRAZIL - With 154 New Confirmed Cases - Dengue Epidemic Fires In Presidente Prudente |
| 2019-05-15 22:21:00 | Detail | Argentina | Resistencia | ARGENTINA - Confirmed Two Cases Of Dengue In Resistencia |
| 2019-05-15 22:18:00 | Detail | Colombia | Bogota | COLOMBIA - Warning In La Guajira For More Than 600 Probable Cases Of Dengue |
| 2019-05-15 22:14:00 | Detail | Brazil | State Of Minas Gerais | BRAZIL - MG Registers An Average Of 54 Thousand New Cases Of Dengue Per Month In 2019 |

FIGURE 1.3 List of country with Dengue hemorrhagic fever [9].

borne viral disease, with a 30-fold increase in global incidence over the past 50 years. The World Health Organization (WHO) estimates that 50–100 million Dengue infections occur each year and that almost half the world's population lives in countries where dengue is endemic. In some countries, the burden of dengue is comparable to that of tuberculosis and other communicable diseases with high disease burdens; unexpected surges in cases and the challenge to health systems of triaging thousands of cases without knowing which severe cases will require hospital care are additional challenges … This Global strategy for dengue prevention and control, 2012–20 aims to address this need.

Dengue morbidity can be reduced by implementing improved outbreak prediction and detection through coordinated epidemiological and entomological surveillance; promoting the principles of integrated vector management and deploying locally-adapted vector control measures including effective urban and household water management. Effective communication can achieve behavioural outcomes that augment prevention programmes.

**FIGURE 1.4** Website of special program for research and training in tropical disease [8].

Research will continue to play an important role in reversing the trend in dengue, a neglected tropical disease, by improving methods and systems for surveillance, prevention and control.

Reversing the trend requires commitments and obligations from partners, organizations and countries, as well as leadership by WHO and increased funding. Dengue prevention and management can now exploit opportunities presented by promising advances in vector control technology interventions, diagnostics, prognostic systems for triage, evidence-based clinical interventions and candidate vaccine developments."

TDR, the Special Program for Research and Training in Tropical Diseases, which is a scientific collaboration supports efforts to combat diseases of poverty has a dedicated section on Dengue virus disease. The program was developed by World Health Organization, and is sponsored by the United Nations Children's Fund, the United Nations Development Program, and the World Bank. On the programs by TDR is the web based tool "Operational Guide: The Early Warning and Response System (EWARS) for Dengue Outbreaks" which acts as a resource for: (i) analysis of historic dengue datasets; (ii) identify appropriate alarm indicators that can predict forthcoming outbreaks; and (iii) use these results and analyses to build an early warning system to detect dengue outbreaks (Fig. 1.4). Another handbook by TDR "Technical handbook for dengue surveillance, dengue outbreak prediction/detection and outbreak response" provides a "model contingency plan" is to assist program managers and planners in developing a Dengue virus illness outbreak response plan through clearly defined and validated alarm signals and

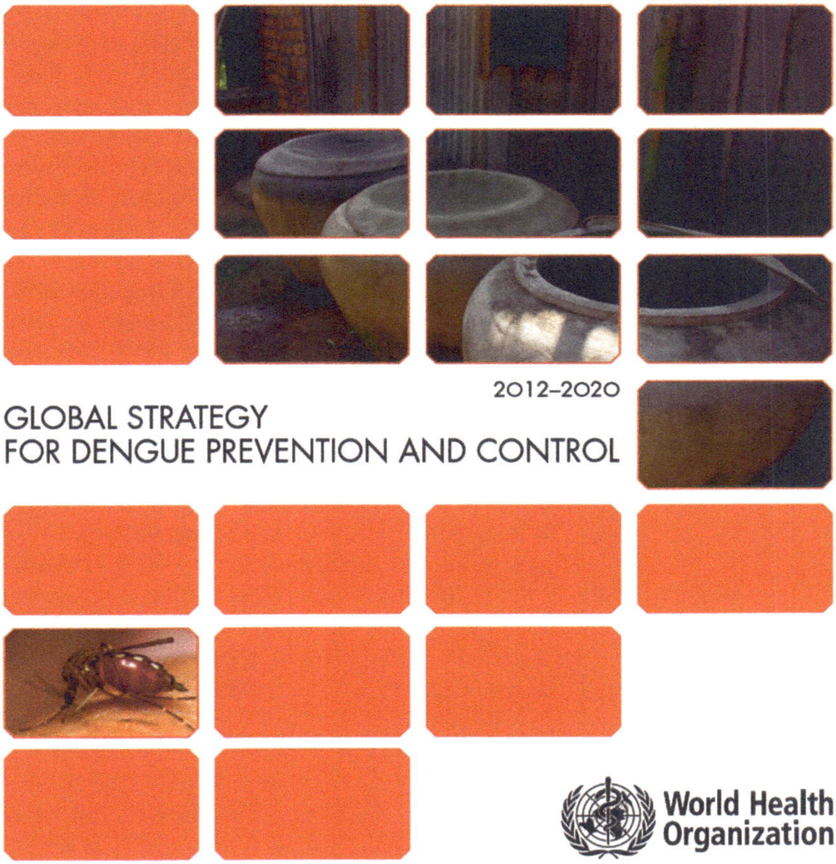


**FIGURE 1.5** World Health organization publication of Dengue viral illness prevention and control [8].

organize an early response "emergency response" once an outbreak has started (Fig. 1.5).

Another controversy is whether donated blood should be screened for presence of Dengue virus [3]. Dengue virus like West Nile virus and Chikungunya virus may be transmitted through blood transfusions. One study tested 15,350 blood donation samples; Dengue virus RNA was detected in 29 samples for a prevalence of 1 per 529 (0.19%). Dengue virus types 1, 2, and 3 with viral titers of 105−109 copies/mL were detected by type specific reverse transcriptase polymerase chain reaction in 12 samples of which all were infectious in mosquito culture [4]. Another study in Portugal found that 43 of 1948 blood donations tested positive for Dengue virus genome (further identified as Dengue virus serotype-1) [5]. Pozzetto et al. reported that only five cases of transfusion-transmitted Dengue viral illness have been confirmed [6]. One patient who received red blood cells containing $10^8$ copies/mL Dengue viral serotype-2, developed Dengue hemorrhagic fever 3 days after transfusion. Both donor and recipient were shown to harbor viruses with the same envelope sequence. Pozzetto et al. [6] recommended the following steps to reduce transfusion related transmission of Dengue viral illness [1] the clinical selection of donors [2]; the implementation of screening tests specific for Dengue virus; and [3] the nonspecific reduction or inactivation of pathogens by the use of physical or chemical treatments applied to blood products. In endemic areas, excluding donors who may be at higher risk of infection is not practical as exposure to mosquito bite is unpredictable. The presence of fever in donors of blood products is a usually a contraindication of blood donation. Gen-Probe Inc. (San Diego, CA, United States) developed a prototype transcription-mediated amplification assay for large-scale screening purpose as in blood donors, with ability to detect approximately 15 copies/mL for each serotype in blood sample. Transcription Mediated Amplification technology simplifies nucleic acid testing by enabling simultaneous detection of multiple viruses in blood sample. Techniques are available to treat blood products in order to inactivate some pathogens. Solvent-detergent treatment to disrupt viral envelopes, exposure to light activated dyes containing phenothiazine like methylene blue, leading to oxidation of guanine present in viral genomes, and nanofiltration to retain viral particles, and photoactivation of compounds by ultraviolet rays. In 2009, the American Association of Blood Banks stratified in four levels (red, orange, yellow and green) the emergent or reemergent infectious agents [7]. The stratification was performed to identify agents that could represent a potential threat to transfusion in North America for the next years. The criteria was based on [1] presence in blood at least for a few hours or days [2]; asymptomatic donors to avoid clinical selection [3]; ability to induce a severe disease; and [4] finally, resist to inactivation by immunity of the donor. Dengue virus was classified in the upper red level, together with *Babesia* sp and the human variant of Creutzfeldt–Jakob disease.

Vaccination for Dengue viral infection has a fascinating story. The Food and Drug Administration on May 1st, 2019 approved the first vaccine "Dengvaxia" from Sanofi Pasteur against Dengue viral illness which was approved by European Commission last year. Dengvaxia is licensed for use in 19 countries, plus the eligible parts of the European Union. The vaccination can increase the risk of severe Dengue viral illness in some individuals. A phenomenon that has resulted in suspension of vaccine and revoking of license in Philippines, one of the countries with largest exposure to the vaccine. Sanofi employees, and several current and former Philippines health officials are being investigated over deaths that could be linked to use of the vaccine. The World Health Organization in September 2018 replaced the 2016 position paper in particular due to new evidence of a greater precipitating severe Dengue viral illness in trial participants who were seronegative compared with those who are seropositive before vaccination. The World Health Organization in September 2018 also recommended "only persons with evidence of a past Dengue infection would be vaccinated (based on an antibody test, or on a documented laboratory confirmed Dengue infection in the past). Only if pre-vaccination screening is not feasible, vaccination without individual screening could be considered in carefully selected areas with recent documentation of seroprevalence rates of at least 80% by the age of 9 years." The position paper also identified the development of highly specific and sensitive rapid tests for determination of Dengue virus serological status as a research priority. Furthermore, research is also needed to evaluate vaccine schedules with fewer doses, and to assess the need for booster doses. The development of safe, effective, and affordable Dengue vaccines for use irrespective of serostatus remains a high priority according to World Health Organization. The Centers for Disease Control and Prevention also is tracking cases in United States as 112 cases have been reported.

## References

[1] Fox news. Philippines declares national emergency after more than 100,000 people contract Dengue fever Fox News Flash top headlines for July 17. Fox News; 2019. Available from: foxnews.com/world/philippines-national-emergency-dengue-fever.

[2] Dayaram S. Philippines declares national alert after 456 die from dengue fever Online: CNN health. 2019. Available from: https://www.cnn.com/2019/07/16/health/philippines-dengue-national-alert-hnk-intl/index.html.

[3] Lanteri MC, Busch MP. Dengue in the context of "safe blood" and global epidemiology: to screen or not to screen? Transfusion 2012;52(8):1634–9.

[4] Stramer SL, Linnen JM, Carrick JM, Foster GA, Krysztof DE, Zou S, et al. Dengue viremia in blood donors identified by RNA and detection of dengue transfusion transmission during the 2007 dengue outbreak in Puerto Rico. Transfusion 2012;52(8):1657–66.

[5] Control ECfDPa. Dengue outbreak in Madeira, Portugal. March 2013. Online.

[6] Pozzetto B, Memmi M, Garraud O. Is transfusion-transmitted dengue fever a potential public health threat? World J Virol 2015;4(2):113–23.

[7] Stramer SL, Hollinger FB, Katz LM, Kleinman S, Metzel PS, Gregory KR, et al. Emerging infectious disease agents and their potential threat to transfusion safety. Transfusion 2009;49(Suppl. 2):1s–29s.

[8] https://www.who.int

[9] https://outbreaks.globalincidentmap.com/

Chapter 2

# Dengue virus disease; the origins

**Omar Saeed[1,2], Ahmer Asif[2,3]**
[1]*Resident Physician, Department of Neurology, University of Tennessee Health Science Center, Memphis, Tennessee, United States;* [2]*Zeenat Qureshi Stroke Institute, St. Cloud, MN, United States;* [3]*Department of Internal Medicine, University of Oklahoma, Oklahoma City, United States*

> *Incredibly, just one mosquito species, Aedes aegypti is responsible for the spread of four known different deadly viral diseases to human beings, yet this mosquito has been allowed to infest densely-populated urban centers.*
>
> T.K. Naliaka.

Another member of the Flaviviridae family with Chikungunya, Yellow fever, and Zika virus is the Dengue virus. This particular virus has seen exponential growth in terms of a global spread over the past several decades. The origins of the word *Dengue* remain unclear; however, the use of the word *Dengue* to describe the disease for the very first time was in Spain in 1801. A most likely theory is that it is derived from the Swahili phrase "Ka-dinga pepo", meaning "cramp-like seizure caused by an evil spirit" [1,2]. The Swahili word "dinga" may possibly have its origin in the Spanish word "Dengue" meaning fastidious or careful, which would describe the gait of a person suffering from the bone pain of Dengue fever. Slaves in the West Indies who contracted Dengue viral illness were said to have the posture and gait of a dandy, and the disease was known as "Dandy Fever." During the 1828 epidemic in Cuba, the illness was first called *dunga,* but later it was changed to Dengue, the name by which we have been addressing this disease ever since.

The first suspected outbreaks of Dengue-like disease were reported in 1635 in Martinique and Guadeloupe and 1699 in Panama [3–5]. However, reports of illnesses compatible with Dengue fever occurred even earlier. The earliest record found to date was in a Chinese "encyclopedia of disease symptoms and remedies," first published during the Chin Dynasty (AD 265–420) and formally edited in AD 610 (Tang Dynasty) and again in 992 during the Northern Sung Dynasty [6]. The Chinese used the name water poison for the disease as it was thought to have a connection with flying insects related to water [30]. Hence, before the 18th century, the Dengue-like disease had a wide geographic distribution where major epidemics occurred widely.

**Dengue Virus Disease. https://doi.org/10.1016/B978-0-12-818270-3.00002-3**

The disease emerged from Africa during the slave trade in the 15th through 19th centuries and spread into the Americas through commercial exchanges in the 17th, 18th, and 19th centuries. The *Aedes aegypti* became highly adapted to the humans and metropolitan environments and sailing ships were the major reason for the spread of this mosquito species throughout the tropics of the world. The infestation first started from the port cities and later moved into the inland. Expansion of the urbanization was solely responsible for it. Over time, *A. aegypti* became intimately associated with the humans, i.e., feeding on them and sharing their living places, making it an efficient epidemic vector of Dengue and Yellow fever viruses [3]. Due to this setting, the major epidemics of Dengue occurred during the 18th till the early 20th centuries (Fig. 2.1).

Early on, after the documentation of mosquitoes as vectors for Yellow fever, many workers in this field suspected mosquitoes to be the vectors for Dengue fever as well. However, in the previrology era due to the slow progress of work and use of human volunteers, its documentation took long. At the beginning of 19th century, the documentation of transmission of Dengue viral illness by mosquitoes was done by Graham in 1903, Bancroft in 1906, and then by Cleland in 1918.

Since after the documentation that mosquitoes transmit the Dengue virus, they were not isolated until the 1940s. During World War II [7—10], in the year 1943, the Dengue virus was isolated for the first time. At that time, Dengue fever was chiefly responsible for illness among Japanese and their allied soldiers in the Asian and Pacific regions. There are four different strains of Dengue virus (DENV1-4) and they are antigenically and phylogenetically different from one another. They were reported for the first time in different regions, as follows:

- DENV1: This strain of Dengue virus was first reported in Japan and French Polynesia in 1943 followed by Hawaii in 1945 [8]. Over time, the reporting of DENV1 in the Asian territories kept on increasing and later in Africa. In the Americas, it was first reported in 1977.
- DENV2: This strain was first identified in 1944 in Papua New Guinea. and Indonesia. Later it was reported in the Philippines, Malaysia and then Thailand. In the Americas, it was first reported in 1953 [11,12].
- DENV3: This strain was reported for the first time in the Philippines and Thailand in 1953. Since then it is been very commonly reported in the Asian territory. The first reports in the Americas were in the year 1963 [13].
- DENV4: This strain was also reported for the first time in the Philippines and Thailand in 1953. Since then it has been reported very commonly in the Asian regions especially Indonesia and Sri Lanka. In the Americas, it was first reported in 1981.

The incidence of a severe form of Dengue fever known as Dengue hemorrhagic fever is not limited to the 20th century. These patients with symptoms of Dengue hemorrhagic fever have been reported since 1780, i.e., in the

FIGURE 2.1 Dengue viral illness occurrence in Africa and the Middle East.

Philadelphia epidemic [14]. Since then, several cases of Dengue hemorrhagic fever have been reported in the form of subsequent epidemics including the ones in Australia (1897), Beirut (1910), Taiwan (1916), and Greece (1928)

[15,16]. Unlike the Dengue fever, these epidemics occurred relatively rarely and had long intervals between them. It made them less important as a long term and continuous public health problem.

## Shifting patterns of dengue fever

Until the 1940s, the Dengue viral illness outbreaks were relatively infrequent but World War II, especially in Southeast Asia, lead to a changed pattern of the disease due to the ecological disturbance. During the years after the war, significant urbanization and economic progress offered ideal conditions for the overwhelming spread of mosquito-borne diseases that in turn led to the start of a global pandemic of Dengue viral illness. Due to the increased epidemic spread and flow of people across the countries, hyperendemicity i.e., cocirculation of multiple Dengue virus strains, developed in the South Asia region leading to the emergence of epidemic Dengue hemorrhagic fever [17]. First known Dengue hemorrhagic fever epidemic occurred in 1953 in Manila and it became more intense by its spread over the next 20–30 years in Southeast Asia. In Asia, the Dengue epidemics geographically stretched to India, Maldives, Pakistan, and Southeast Asian countries east to China [3,17]. Numerous island countries of the Central and South Pacific regions (New Caledonia, Tahiti, Cook Islands, Palau, Niue, Yap, and Vanuatu) have also experienced a number of minor and major Dengue hemorrhagic fever outbreaks [18] (Fig. 2.2).

In the Americas, the changes in the epidemiology were the most histrionic. During the 1960s and 1970s, the epidemics of Dengue viral illness were rare in the American territories due to the eradication of *A. aegypti* from the Southern and Central parts of America [19–21]. In the early 1970s, the eradication program was stopped which led to the reinvasion of the infection in the countries from where it was already eradicated [20,21]. Until the 1990s, the *A. aegypti* reinfested that geographic distribution again. During this time, the American region was facing the major Dengue viral illness epidemics that had been free of the disease for the past 100 years [19,20,22]. Similar to the Southeast Asian region, the development of hyperendemicity due to the peaked epidemic activity in the American region also led to the emergence of epidemic Dengue hemorrhagic fever. Since the year 1981–2015, Dengue hemorrhagic fever was confirmed on laboratory reports in 24 American countries (Fig. 2.3) [22,23].

In Africa, the sporadic cases of Dengue hemorrhagic fever occurred more commonly than having major epidemics. This is due to the remarkable increase in the Dengue fever epidemic in the past 25 years in this region leading to severe disease. Until the 1980s, very little was known about the spread of Dengue fever disease viruses in Africa. Since then, major Dengue fever epidemics have occurred in both the Western and Eastern parts of Africa [18,24], which involved all four viral strains. In the 1990s, these outbreaks were more

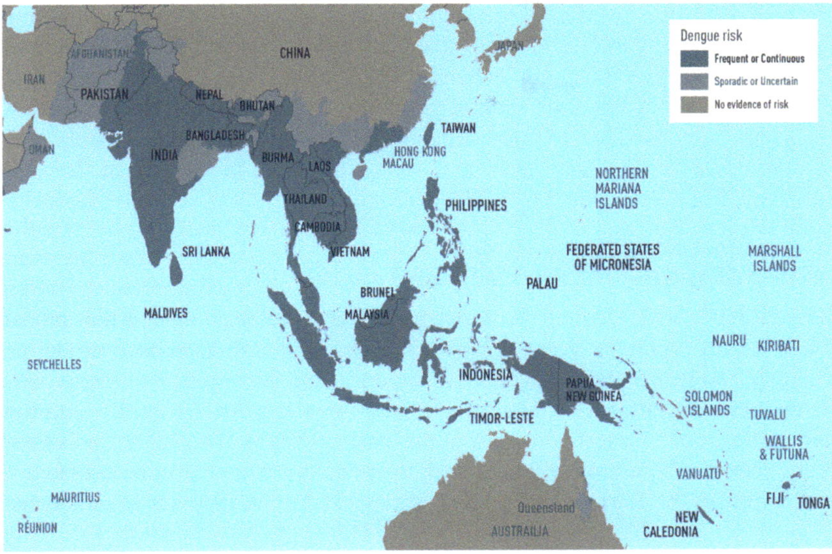


FIGURE 2.2 Dengue viral illness occurrence in Asia and Oceania.

common in the Middle East and East Africa, having the major ones in Djibouti in 1991 and in Jeddah, Kingdom of Saudi Arabia in 1994 [18].

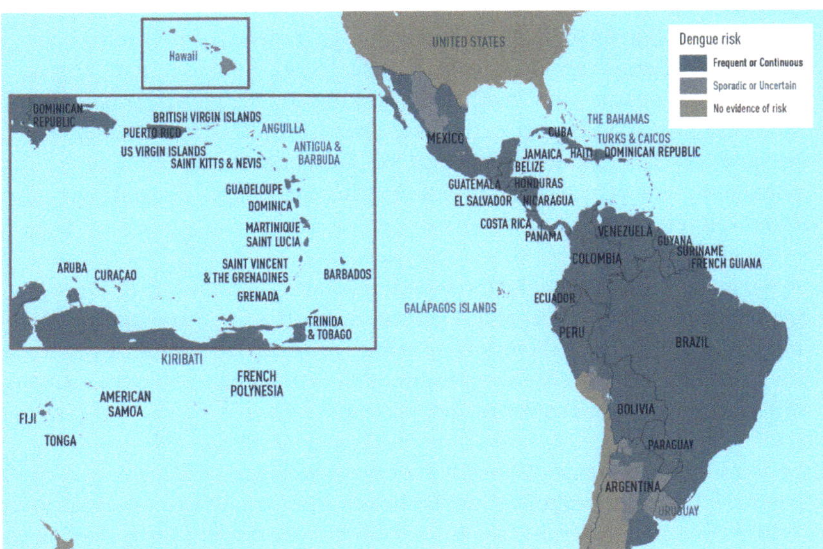


FIGURE 2.3 Dengue viral illness occurrence in the Americas and Caribbean.

In 1997, the *A. aegypti* mosquitoes and Dengue viruses had global dissemination in tropics and more than 2.5 billion humans dwell in regions where the Dengue fever is endemic [23,25,26]. Presently, this virus is responsible for causing more morbidity and mortality than any other arbovirus illness in humans. Due to these epidemics, every year approximately 100 million cases of Dengue fever and many hundred thousands of Dengue hemorrhagic fever occur [23,25,27].

## Factors responsible for increased incidence

Dengue fever and Dengue hemorrhagic fever outbreaks have been global public health problems over the past 17 years. Although a number of factors are responsible for the significant resurgence and emergence of these outbreaks but still the precise determination of these factors is complex and not well understood. Nevertheless, over the past 50 years [18,23,24] this resurgence seems to be narrowly linked with the demographic and societal changes. Following are the four major factors responsible for the increase in incidence:

1) One of the major factors has been the extraordinary growth of the global population. This population growth has been the main driving force for the uncontrolled and unplanned urbanization, particularly in tropical countries. In this regard, the substandard housing, overcrowding of cities, and decline in sewer, water, and waste management systems are aiding to provide an ideal environment for the increased transmission of vector-borne diseases.

2) A second major factor is the deficiency of an effective mosquito control program in Dengue-endemic areas [19,20,23,24]. Since the past 25 years, spraying the spaces with insecticides to kill the mosquitoes is being used but it has proven to be ineffective over time [20,28,29]. In addition to that, because of the augmented amount of mosquito larval habitations, the population density and terrestrial distribution of *A. aegypti* has also increased, especially in the tropics.

3) Another most important factor contributing to the increase in the emergence of Dengue fever and Dengue hemorrhagic fever outbreaks is increased travel by airplanes. Air travel provides an ideal way for the transportation of viruses like dengue and other urban pathogens among different population centers of the world [23,24].

4) Fourth important factor responsible for the resurgence of Dengue outbreaks has been the deterioration of public health infrastructures in the underdeveloped countries over the last 30 years. Scarcity of resources has led to an overwhelming shortage of qualified physicians who can propose and

develop effective prevention and control programs for the mosquito and other vector-borne diseases.

In summary, the societal and demographic changes, lack of effective mosquito control programs, scarce resources for the vector-borne disease prevention and control, and alterations in the public health program have all led to the increased Dengue epidemic activity, the hyperendemicity development, and the incidence of Dengue hemorrhagic fever epidemic. As of 2019, the global burden of Dengue viral illness is truly immense affecting approximately 2.5 billion or 40% of the world's population being endemic in over 100 countries including Asia, Pacific, the Americas, Africa, and Caribbean. According to World Health Organization, there are anywhere from 50 to 100 million cases every year with staggering 22,000 deaths due to the Dengue viral illness.

## References

[1] Christie J. On epidemics of dengue fever: their diffusion and etiology. Glasgow Med J 1881;16(3):161.

[2] Christie J. Remarks on "Kidinga Pepo": a peculiar form of exanthematous disease. British Med J 1872;1(596):577.

[3] Gubler DJ. Dengue and dengue hemorrhagic: its history and resurgence as a global public health problem. In: Dengue and dengue hemorrhagic fever. London: CAB International; 1997. p. 1–22.

[4] McSherry JA. Some medical aspects of the Darien scheme: was it dengue? Scott Med J 1982;27(2):183–4.

[5] Halstead S. Dengue: overview and history. In: Tropical medicine: Science and practice. London: Imperial College Press; 2008.

[6] Nobuchi H. The symptoms of a dengue-like illness recorded in a Chinese medical encyclopedia. 1979. p. 422–5.

[7] Kimura R, Hotta S. Studies on dengue: anti-dengue active immunization experiments in mice. Jpn J Bacteriol 1944;1:96–9.

[8] Hotta S. Experimental studies on dengue: I. Isolation, identification and modification of the virus. J Infect Dis 1952;90(1):1–9.

[9] Sabin AB. Research on dengue during world war II. Am J Trop Med Hyg 1952;1(1):30–50.

[10] Sabin AB, Schlesinger RW. Production of immunity to dengue with virus modified by propagation in mice. Science 1945;101(2634):640–2.

[11] Rico-Hesse R, Harrison LM, Salas RA, Tovar D, Nisalak A, Ramos C, et al. Origins of dengue Type 2 viruses associated with increased pathogenicity in the Americas. Virology 1997;230(2):244–51.

[12] Cologna R, Armstrong PM, Rico-Hesse R. Selection for virulent dengue viruses occurs in humans and mosquitoes. J Virol 2005;79(2):853.

[13] Messer WB, Gubler DJ, Harris E, Sivananthan K, De Silva AM. Emergence and global spread of a dengue serotype 3, subtype III virus. Emerg Infect Dis 2003;9(7):800.

[14] Rush B. An account of the bilious remitting fever: as it appeared in philadelphia, in the summer and autumn of the year 1780. Am J Med 1951;11(5):546–50.

[15] Copanaris P. L'Epidémie de dengue en Grèce au cours de l'été 1928. In: Par P. Copanaris: Office international d'hygiène publique; 1928.
[16] Akashi K. A dengue epidemic in the Tainan District of Taiwan in 1931. Taiwan No Ikai 1932;31:767.
[17] Halstead SB. Dengue haemorrhagic fever—a public health problem and a field for research. Bull World Health Organ 1980;58(1):1.
[18] Gubler DJ. Dengue and dengue hemorrhagic fever: its history and resurgence as a global public health problem. In: Dengue and dengue hemorrhagic fever; 1997.
[19] Gubler D. Epidemiology of arthropod-borne viral diseases. Boca Raton, USA: CRC Press Inc; 1988.
[20] Gubler DJ. Aedes aegypti and Aedes aegypti-borne disease control in the 1990s: top down or bottom up. Am J Trop Med Hyg 1989;40(6):571−8.
[21] Pinheiro FP. El dengue en las Américas: 1980−1987. 1989.
[22] Pinheiro FP, Corber SJ. Global situation of dengue and dengue haemorrhagic fever, and its emergence in the Americas. World Health Stat Q 1997;50:161−9.
[23] Gubler D. The global pandemic of dengue/dengue haemorrhagic fever: current status and prospects for the future. Ann Acad Med Singapore 1998;27(2):227−34.
[24] Gubler D, Trent D. Emergence of epidemic dengue/dengue hemorrhagic fever as a public health problem in the Americas. Infect Agents Dis 1993;2(6):383−93.
[25] Gubler DJ, Clark GG. Dengue/dengue hemorrhagic fever: the emergence of a global health problem. Emerg Infect Dis 1995;1(2):55.
[26] Halstead SB. Pathogenesis of dengue: challenges to molecular biology. Science 1988;239(4839):476−81.
[27] Monath TP. Dengue: the risk to developed and developing countries. Proc Natl Acad Sci 1994;91(7):2395−400.
[28] Newton EA, Reiter P. A model of the transmission of dengue fever with an evaluation of the impact of ultra-low volume (ULV) insecticide applications on dengue epidemics. Am J Trop Med Hyg 1992;47(6):709−20.
[29] Reiter P, Gubler DJ. Surveillance and control of urban dengue vectors. New York: Dengue and dengue hemorrhagic fever CAB International; 1997. p. 425−62.
[30] Bock G, Goode J, editors. New treatment strategies for dengue and other flaviviral Diseases: Novartis Foundation Symposium 277, vol. 277. First published: 25 August 2006. History Page 3−4.

## Further reading

[31] Gubler DJ. Dengue and dengue hemorrhagic fever. Semin Pediatr Infect Dis 1997;8(1):3−9.

Chapter 3

# Dengue virus infection outbreak: comparison with other viral infection outbreak

**Mohammad Rauf A. Chaudhry**
*Resident Physician, Texas Tech University Health Sciences Center, El Paso, TX, United States; Zeenat Qureshi Stroke Institute, St. Cloud, MN, United States*

Dengue virus infection is one of the most common arthropod-borne viral disease causing about 50−100 million infections per year [1]. About 40% world population live in areas with high risk of Dengue virus transmission. There are about 100 countries in Asia, the Pacific, the Americas, Africa, and the Caribbean where Dengue virus infection is endemic [2]. Dengue virus infection is caused by any of the four narrowly related serotypes: Dengue viruses 1−4. Infection with one serotype does not provide immunity against other serotypes. In fact it increases the risk for Dengue hemorrhagic fever and Dengue shock syndrome [3]. The four serotypes originated in monkeys and independently made a cross over to humans in Africa or South East Asia about 100−800 years ago [4].

Dengue virus infection can be asymptomatic or a self-limited, varying in severity, classical form is characterized by high fever, headache, stomach ache, rash, myalgia, and arthralgia. Dengue hemorrhagic fever and Dengue shock syndrome are severe forms of Dengue virus infection, accompanied by thrombocytopenia, vascular leakage, and hypotension [1]. Mechanisms underlying the severe form of disease are still not well understood despite the intensive research. The lack of understanding is partly due to lack of appropriate animal models of infection and disease. Due to lack of vaccine and antiviral drugs, only control measure is limiting the Aedes mosquito vectors spread [1].

## Epidemic versus pandemic

Epidemic and pandemic are primarily different in terms of spread of contagious, infectious, or viral illness. Epidemic is limited to one specific region and

**Dengue Virus Disease. https://doi.org/10.1016/B978-0-12-818270-3.00003-5**

is an event in which a disease is actively spreading while pandemic describe a disease affecting the whole country or the entire world. In simple words, when epidemics falls short in describing the scale of a problem, it is better to use pandemic [5].

## Common features of epidemics

According to our previous work on Ebola virus infection epidemic by Qureshi et al. [6], there are several factors that can start an epidemic listed below:

1 Disasters (e.g., wars, famine, floods, and earthquakes)
2 Temporary population settlements
3 Preexisting diseases in the population
4 Ecological changes like floods and cyclones
5 Resistance potential of the host (i.e., nutritional and immunization status of the host)
6 Damage to public utility and interruption of public health services

Qureshi et al. mentioned that there are three patterns of disease continuity:

1 Saw tooth pattern
2 Tooth necklace pattern
3 Tooth eruption pattern

## Saw tooth pattern

It represents an intermittent outbreak of a disease that recedes in intensity, but the disease is not eradicated from the population. The smallpox epidemics in Africa during 1920–1950s would be an example of such a pattern Figs. 3.1–3.3.

## Tooth necklace pattern

It constitutes where the disease is eradicated from the population, but pathogen species is kept alive under controlled circumstances for preparation of vaccines and biological studies. While the escape of pathogen from confinements of laboratories has been the subject of numerous conspiracy theories, vaccination with live attenuated viruses is more likely to be the string to maintain the continuity.

## Tooth eruption pattern

It constitutes where, like the tooth hidden within the gums and emerging independent of other teeth, the pathogen emerges and is exterminated without

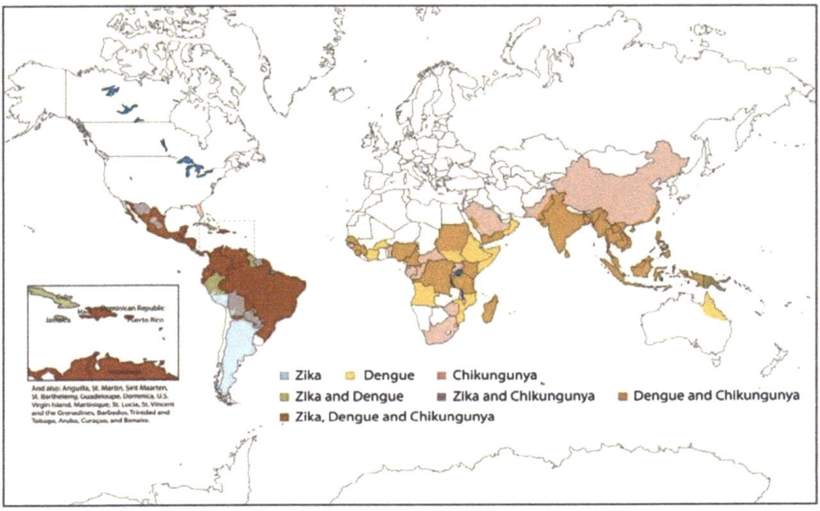


**FIGURE 3.1** Map showing the estimated global distribution of Dengue, Zika, and Chikungunya [9].

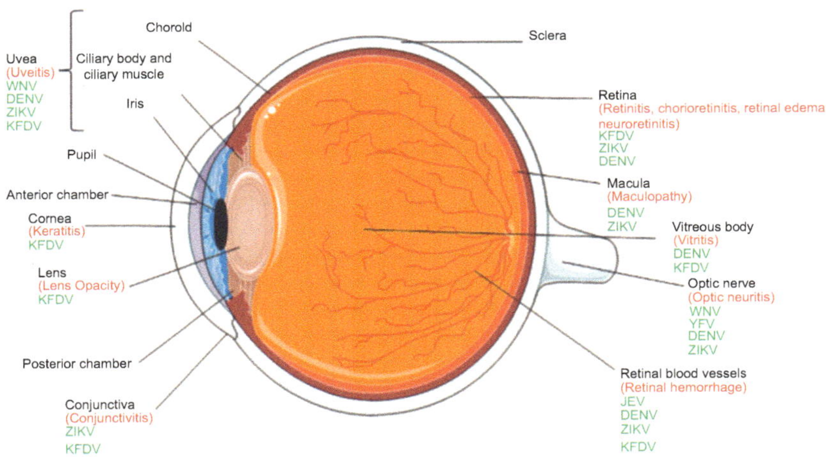


**FIGURE 3.2** Eye anatomy and ocular complications caused by flaviviruses. Various components of the human eye are labeled in black. The flaviviruses responsible for causing ocular manifestations are shown in green whereas specific ocular tissue pathology is highlighted in red [11].

any relation to previous occurrences. The Dengue virus infection epidemic follows the "tooth eruption" pattern.

## Why epidemics die their deaths?

It is generally believed that measures like such as vaccination of at-risk individuals, quarantine of diseased persons, and acute and timely treatment help to control all the epidemics. However, facts do not support this conclusion. In

| Ocular Complication | ZIKV | DENV | JEV | WNV | YFV | KFDV |
|---|---|---|---|---|---|---|
| Conjunctivitis/keratitis | + | + | − | − | − | + |
| Macular mottling | + | + | − | − | − | − |
| Chorioretinal atrophy | + focal pigmentary clumping | + | − | − | − | + |
| Optic nerve abnormalities | + | + | − | + | + | − |
| Cataract | Hypoplasia, cupping, pallor | − | − | − | − | − |
| Microphthalmia | + | − | − | − | − | − |
| Iris coloboma | + | − | − | − | − | − |
| Uveitis | + | + | − | + | − | + |
| Chorioretinitis | + | + | − | − | − | + |
| Retinal hemorrhage | + | + | + | − | − | + |

ZIKV—Zika virus; DENV—Dengue virus; JEV—Japanese Encephalitis virus; WNV—West Nile virus; YFV—Yellow Fever virus; KDFV—Kyasanur Forest Disease virus.

**FIGURE 3.3** Comparison of the major ocular findings among the different flaviviruses infections [11].

fact, the largest epidemics, such as the Peloponnesian War Pestilence, Antonine Plague, Plague of Justinian, Black Death of the 14th century, and Spanish flu, came to an end without widespread use of any of these strategies mentioned above.

Qureshi et al. [6], came up with three theories for spontaneous remission of epidemics which are mentioned below:

1. There are two types of people within the exposed population: some more vulnerable and some more resistant. The people who may be resistant to the disease may be so because of previous exposure to viruses with similar structures, resulting in the development of immune responses that are adequate for multiple pathogens. They might also be resistant due to superior health, including age, nutritional status, and occupational advantages. The virus might eventually be faced with a population that is completely resistant to the infection.
2. Changing environment within habitats that are not conducive to the survival or propagation of viruses or other pathogens. Weather changes, including temperature and humidity fluctuations, may significantly influence the survival or propagation of a virus outside the body. Elimination of reservoirs that carry pathogens including animals, insects, food, or water, by chance or design, may disrupt the cycle of propagation. Such elimination of infection is less likely to occur within an epidemic because of diverse factors and geographical areas involved.
3. The most likely explanation is the "Sand Filter Theory," a term coined by Dr. Qureshi. This theory reflects the similarity between retention of particulate matter during filtration based on density of sand particles, which can be compared to pathogens within a population based on population density.

Most epidemics are composed of diseases that require close contact between diseased and healthy individuals for continued propagation of pathogens. Unlike natural disasters, such as hurricanes, floods, volcanoes, and changes in climate that exist independent of population density, epidemics depend upon population density, a feature shared with reproduction rates, migrations, and predation. After population density reduces below a critical limit, such contact may not be available enough for continued propagation of pathogens.

## Review of the factors modulating Dengue virus infection transmission

Understanding of factors, which play important role in Dengue virus infection transmission, is very important for planning more effective strategies for prevention and control of this disease [7].

## Relations among rainfall, vector density, and Dengue virus infection incidence

A positive association between rainfall or larval density and Dengue viral illness incidence has been reported by few studies but it only hold true for countries just above and below the equator. Generalization of this association to drier regions has been flawed. For example, Dengue virus infection epidemics have been recorded in regions where rainfall or larval indices were unusually low or where due to availability of piped water, water jars/wells were rare. In northeastern Thailand in 1987, Dengue hemorrhagic fever epidemic happened in dry and hot season and was over before the rainy season. It is important to note that Dengue hemorrhagic fever incidence was highest in that region of the country [7].

## Temperature

Blanc and Caminopetros reported that *Aedes aegypti* mosquitoes can only transmit the Dengue virus at temperature above 20°C. For example, the Dengue virus infection epidemics ceased when the temperature dropped to 14−15°C in winter. Global warming may affect this seasonal variation and vector distribution. As *Aedes aegypti* mosquitoes infect mostly domesticated species, as a result outdoor temperature may not always affect their distribution. That is why it was not surprising to have Dengue virus infection outbreak in Mexico at an altitude of 1700 m [7].

## Minimum threshold of vector density for Dengue virus infection transmission

Though minimum vector density below which the Dengue virus infection transmission ceases is possible but, minimum threshold has not been defined yet. The validity of using the minimum vector density to access the success of programs directed toward Dengue virus infection has been questioned [7].

## Vector movement

### Flight range

The flight range for *Aedes aegypti* mosquito has been variable from as long as 2.5 km over 24 h, to as short as 25 m in desert and urban environments, respectively The possible reasons for this variation could be finding suitable sources of food such as nectar, human hosts, an oviposition site, or resting place along the flight path. In one study conducted in an African village, the majority of marked mosquitos remained in the house when they were recaptured. In addition, the house entering behavior of mosquito is genetically controlled [7].

### Transport of vectors

Three important modes of vector transmission are: water, land, and air. Before, the advent of air travel, ships were considered to be the principal source for transmission of Dengue virus infection from Africa to Asia and Americas. Latter on development of efficient highways and other means of ground transportation lead to infestation by the *Aedes aegypti* mosquito in towns located along the roads and railways. Similarly, an extensive list of records showing aircraft bringing *Aedes aegypti* mosquitoes to Dengue-free countries is available [7].

## Mosquito-related factors

Density (number) of Dengue virus—infected adult female mosquitos per residence is an important factor in transmission but is not the only factor. For example, in an isolated Dengue virus infection outbreak in a school in Malaysia, only three *A. aegypti* female mosquitoes were collected from a hostel when there 20 students were infected. Other factors include proportion of engorged females mosquitoes per residence, number of virus infected females, and multiple feedings (bites). Multiple feeding per mosquito is considered to be an important factor for exponential spread of Dengue virus infection epidemic, but it is also not the only factor. It depends if the first human bitten had neutralizing antibody which could result in neutralizing of virus from a viremic person to a noninfected person, provided the second bite is within 6 h after first bite. Other factors include basic reproductive rate, extrinsic incubation period (the period between an insect's feeding on a viremic person and it becoming infective), and amount of virus in infective vector and amount injected by bite [7].

### Human factors

More than one person per household increases the chance of transmission of Dengue virus infection epidemic. Transmission occurs if same mosquito infects other susceptible member of the household or an uninfected female mosquito feeds on first victim who is still in viremic stage of illness. Thus multiple infections in a single household accelerate the spread of infection in the community. Vector-infested places: schools, commercial establishments, churches or temples, offices, military bases, factories, hospitals, prisons, and theaters facilitate the Dengue virus infection spread [7].

### Herd immunity

With respect to role of protective antibody, it is difficult to determine the exact levels of immunity required against a specific serotype for all age groups in secondary infection. In one study in Singapore, despite the low levels of mosquito density, low levels of herd immunity was likely responsible for continuous spread of Dengue hemorrhagic fever in children under age 10 years. Thus it was reported that intense vector control program in Singapore brought the opposite effects with increase in outbreaks of Dengue fever as a result of declining herd immunity. For some viral diseases such as measles and rubella, levels of immunization required for disease prevention range from 84 to 96%. Thus knowing levels of herd immunity required for Dengue virus infection prevention is vitally important [7].

## Breast milk as a possible route of Dengue virus infection vertical transmission

Barthel et al. [8]. reported breast feeding as possible route for Dengue virus infection transmission. They reported a case report where patient presented to hospital in preterm labor and gave birth to a premature but healthy baby. On day 2 after birth, infant was fed on expressed breast milk after which both mother and baby experienced nonsevere acute Dengue infection with fever and severe thrombocytopenia but no signs of hemorrhage or plasma leakage. Dengue virus was tested in breast milk and as result breast feeding was stopped on day 4. The viremic period was prolonged ($\geq$10 days) in newborn which could be related to child's prematurity. Breast feeding transmission has been reported in other flaviviruses such as West Nile virus and Yellow fever virus. Therefore, recommendations have been made to stop breast feeding during the acute viremic phase after using a live attenuated virus vaccine [8].

## Dengue, Zika, and Chikungunya viruses: emerging arboviruses in the new world

Dengue, Chikungunya, and Zika viruses are all three arboviruses, which in recent years have expanded across the globe with large outbreaks in Western Hemisphere territories in close proximity to the United States. The increase in globalization led to spread of these infections to populations with no native immunity. These viruses spread to human through bite of Aedes species mosquito (*Ae. aegypti* and *Ae. Albopictus*). These mosquitoes are aggressive bitters during the day time but can also bite at night too. These mosquitos lay eggs in standing water in things like buckets, bowls, animal dishes, flower pots, and vases [9].

### Zika virus

Zika virus is named after an Ugandan forest in which it was first discovered and is closely related to Dengue virus. For decades, it was of a little concern for clinicians, until recently when a correlation between Zika virus infection and fetal microcephaly was discovered in 2016 and World Health Organization officially declared it a "Public Health Emergency of International Concern." First large outbreak of Zika virus infection occurred in 2007 in Yap, a small island in Micronesia. About 73% population of the island got infected during this outbreak. Subsequent outbreaks occurred across the Pacific Islands until 2015 when Brazil reported the first case of Zika virus infection in the America. Other possible mechanisms other than mosquito biting are sexual transmission and blood borne transmission likely during the viremia stage. Zika virus has also been isolated from urine, saliva, and breast milk of infected individuals but no transmission has been documented from these sources yet. Incubation period for Zika virus about 2 weeks. During viremia, a mild disease consisting of fever, nonpurulent conjunctivitis, a maculopapular rash, arthritis/arthralgias, headache, and vomiting occurs. Zika virus infection has not been shown to cause the severe capillary leak syndrome or hemorrhagic fever. Almost 80% infections are asymptomatic. Since symptoms are clinically indistinguishable from Dengue virus infection initially, aspirin, nonsteroidal antiinflammatory drugs, and steroids should be avoided as they may increase the severe hemorrhage in cases of Dengue infection. Due to potential risk of fetal birth defects in pregnant women due to Zika virus infection, it is strongly recommended by Centers for Disease Control and Prevention to avoid travel to endemic areas. If the travel cannot be avoided, strict mosquito protection measures should be taken. Pregnant women with Zika virus infection should undergo serial ultrasounds every 3–4 weeks throughout the pregnancy. In addition, there is also a concern for possible correlation between Zika virus infection and Guillain–Barre syndrome. Some occurrences have been reported by several countries in the western Pacific and Americas [9].

### Chikungunya virus

Chikungunya virus was first isolated 1953 from a febrile patient in Tanzania. In 2004, large outbreaks occurred in throughout Africa and Asia. Chikungunya, an alphavirus of the Togaviridae family is a mosquito-spread virus and cause the symptoms similar to Dengue and Zika virus infections. Viremia and other symptoms occur after an incubation period of 1–12 days (typically 3–7 days). Arthralgia almost occurs in all the cases with common joints involved are ankles, wrists, and fingers. Arthralgia is worse in the morning, improves with mild exercise but worsens with strenuous exercise. Similar to Dengue and Zika viral infections, management of Chikungunya viral infections is supportive. Acetaminophen is preferred for pain and fever control over nonsteroidal antiinflammatory drug due to risk of possible misdiagnosis of Dengue virus infection [9].

## Dengue and other emerging flaviviruses

More than 70 flaviviruses have been identified [10], almost half of them cause disease in humans and only few are of major importance. There are three clinical syndromes caused by the flaviviruses: fever-arthralgia-rash, viral hemorrhagic fever with or without hepatitis, or central nervous system diseases. There are no antiviral drugs against the flaviviruses but vaccines do exist against a few of them. Mosquito-borne viruses tend to occur in warm while the tick-borne viruses in cooler climates. Examples are mosquito-borne Japanese encephalitis virus occurs in southern and eastern Asia but tick-borne encephalitis virus occurs in Europe and Commonwealth of Independent States (the former Soviet Union). In addition, mosquito-borne viruses have shorter life cycles due to the fact that mosquito have shorter life cycles than ticks. As a result, mosquito-borne viruses are evolving rapidly to fill in the ecological niches in new geographical areas. Flaviviruses other than Dengue and yellow fever viruses are enzootic meaning they use birds or small mammals as the natural hosts as these animals have high reproductive rates which provide them a ready supply of immunologically naïve hosts. Humans get infected from these enzootic viruses when they come in close proximity with the mosquitoes' natural cycle. As humans do not produce high viremia, therefore act as a dead end hosts for them. Most of the flavivirus infections produce a mild disease with the exception of few diseases [10].

## Yellow fever virus

Epidemics compatible with yellow fever virus have been described in West Indies, Central and South America and, West Africa since 15th century. In the early 1900s, a team demonstrated that yellow fever was caused by a filterable agent (a virus), transmitted by mosquitoes. Yellow fever occurs in jungle and urban cycles in West and South America, transmitted to forestry and agricultural workers when bitten by mosquitoes. Due to high and prolonged viremias in humans, infected individuals will carry the disease to populated areas where Aedes mosquitoes transmit the virus to cause "urban Yellow Fever." Yellow fever is characterized with high grade fever, headache, back and muscle ache nausea, and vomiting with more disease as liver failure occurs, causing the mild jaundice which gives disease its name. Yellow fever vaccine was one of the first live attenuated vaccines with conferring immunity up to 10 years or more. Other effective measures included removal of breeding sites, treatment of stored water, and ultralow volumes spraying [10].

## Japanese encephalitis

Japanese encephalitis is the most important viral encephalitis worldwide with 50,000 cases and 10,000–15,000 deaths. The disease was recognized since epidemics of encephalitis in Japan in 1870s onwards and virus was isolated from a fatal case in 1930s. Virus gets transmitted between birds and animals by Culex mosquitoes, especially *Culex tritaeniorhynchus* and humans get infected due to close proximity with this enzootic cycle. Serological studies have shown that almost all individuals living in rural Asia get infected with Japanese encephalitis during the childhood but only 1 in 300 become symptomatic. Japanese encephalitis virus infections cause variable clinical presentation ranging from headache, cough, and coryza but in some cases these febrile prodromes are followed by coma and convulsions. About 20%–30% patients die and more than half of the survivors are left with severe neuropsychiatric sequelae. Cerebrospinal fluid analysis will likely show lymphocytic pleocytosis but can be normal and computer tomography and magnetic resonance imaging shows damage in thalamus, basal ganglia. Japanese encephalitis vaccine is available for 30 years now and strongly recommended for residents of endemic areas and travelers planning to visit the endemic areas for more than 3 weeks. Other important preventive measures include insect repellents; bed nets to avoid mosquitoes biting; and keeping pigs, chickens, and other potential animals away from human dwellings [10].

## West Nile encephalitis

West Nile virus, first isolated in 1930s, was considered benign until recently and is mostly found in much of Africa, much of Asia, Southern Europe, and recently in North America. Like Japanese encephalitis, West Nile encephalitis transmitted by Culex mosquitoes and humans are dead hosts. Virus will cause fever-arthralgia-rash syndrome mostly with conjunctivitis and lymphadenopathy. Clinically, West Nile encephalitis is similar to Japanese encephalitis in terms of causing coma, convulsions, and mixture of causing upper and lower motor neuron signs: flaccid paralysis of limb and respiratory muscles. Meningoencephalitis is considered a rare complication of West Nile virus infection. A large outbreak occurred in Romania, in 1960s in which about 600 people had the neurological disease but no one with rash or lymphadenopathy. Poor plumbing and sewerage under the apartments was considered to cause the outbreak as it lead to explosive increase in population of *Culex pipiens*. All age of people were affected but the attack rate and mortality was higher in older adults [10].

## Other mosquito-borne flaviviruses

Other important mosquito-borne flaviviruses include Murray Valley encephalitis virus, Kunjin virus, Saint Louis encephalitis virus, and Rocio virus. Murray Valley encephalitis virus was the cause of polio like encephalomyelitis in Australia in 20th century. The virus gets transmitted by *Culex annulirostris*, water birds, cattle, and some marsupials. Kunjin virus is widely distributed geographically but cause a milder disease. St Louis encephalitis virus was an important neurotropic flavivirus in United States with first epidemic reported in 1930s. Though most of the areas of United States, Canada, and Mexico are affected by this disease at some point in time, sporadic cases still occur. Rocio virus was first isolated from the brain of a fatal case from an outbreak in Sao Paulo State of Brazil in 1975 and is transmitted by Psorophora mosquitoes [10].

## Tick-borne flaviviruses

Tick-borne flaviviruses are less important in humans compared to animals as ticks preferentially feed on animals compared to humans. Viruses are transmitted by Ixodidae (hard ticks): Dermacentor and Hemophysalis species. Other important routes of transmission are ingestion of infected milk, and direct transmission from infected animal carcasses. If no host is available, ticks can act as virus reservoirs for months and years. Tick-borne encephalitis mostly circulate among the small wild animals like rodents and transmitted to human by ingestion of goat milk in addition to tick bites. Up to 70% of the patients remember the tick bite with tick-borne encephalitis presenting 1–2 weeks after the high grade fever, headache, malaise, and myalgia. In latter cases, it can progress to flaccid paralysis of upper limb and shoulder girdle. Respiratory muscles and bulbar (brain stem) involvement can lead to respiratory failure and death [10].

## Ocular manifestations of emerging flaviviruses and the blood-retinal barrier

Flaviviruses, despite causing many systemic complications, has been documented to cause multiple ocular abnormalities such as conjunctivitis, retinal hemorrhages, chorioretinal atrophy, posterior uveitis, optic neuritis, and maculopathies [11].

Eyes are protected from the systemic infections by presence of blood–retinal barriers. Flaviviruses modulate the retinal innate response and penetrates the blood–retinal barriers to cause the ocular pathologies [11].

## Seroepidemiology of Dengue, Zika, and Yellow Fever viruses among children in the democratic republic of the Congo

The public health importance of arthropod-borne viruses is growing tremendously as they cause millions of infections in humans annually with physical manifestations ranging from birth defects, hemorrhage, shock, encephalitis, and even death. Dengue fever is causing about 400 hundred millions infections annually. In 2015, Zika virus infection in Latin America became an international public health emergency due to its adverse effects on developing fetus when expecting mothers were infected. The epidemiology of Dengue virus infection and Zika virus infection in Asia and America better described compared with Africa. Finding Dengue and Zika virus infections in travelers returning from Africa suggests that prevalence of these viral infections in African population is largely underestimated. A recent outbreak of yellow fever virus in 2016 in Angola and surrounding countries despite the existence of an effective vaccine since the 1930s represents a constant threat for yellow fever virus epidemic [12]. At least 42 deaths were reported in Democratic Republic of the Congo when the Angolan outbreak crossed country orders. Risk of Yellow fever virus outbreaks in the Democratic Republic of the Congo and lack of data on other flavivirus infections such as Dengue virus and Zika virus prompted Willcox et al. [12] to conduct a seroepidemiological survey for these three flaviviruses. They found that children despite the documented proof of receiving the Yellow fever virus vaccine, failed to show the evidence of seroconversion. Evidence of low rate of seroconversion among children has been previously reported but it consistently reported to be more than 80%. The other possible reason could be that majority of children had administration of Measles vaccine with Yellow fever virus vaccine. It has been demonstrated that coadministration of measles vaccine with yellow fever vaccine decreases its immunogenicity [12]. Other reasons include malnutrition in children and insufficiency of sensitivity of enzyme-linked immunosorbent assay and neutralization assays to detect the low levels of antibodies that are sufficient for protection. Despite the possibility of above mentioned reasons, the fact remains that Democratic Republic of the Congo continues to experience major yellow fever outbreaks which questions the effectiveness of efforts to integrate yellow fever vaccine into routine childhood vaccination programs. In addition, the study found that Dengue, Zika, and yellow fever viruses are circulating in the Democratic Republic of the Congo. Recommendations were made to conduct more studies to explore the detailed reasons for low rates of seroconversion observed in vaccinated Congolese children and considering flavivirus infection as an important etiology of acute febrile illness especially in patients who test negative for malaria [12].

## Viremia and clinical presentation in Nicaraguan patients infected with Zika virus, Chikungunya virus, and Dengue virus

All the three viruses cocirculate in Nicaragua. Study was conducted to compare the clinical presentation and quantify the levels of viremia. Out of 263 patients tested positive: 192 tested positive for a single virus infection, 71 for two, and 2 for all the three viral infections (coinfections). Viremia levels were lower in Zika virus infections compared with Chikungunya virus or Dengue virus. Zika virus infected patients were likely to develop rash but less likely to be febrile or hospitalized compared to chikungunya virus or Dengue virus—infected patients. Due to lot of similarity in clinical presentation, it becomes difficult to make accurate clinical diagnosis where the patient may be infected with any of three viral infections. This supports the use of testing protocol for sensitive, accurate, multiplex diagnostics for clinical care, disease research, and epidemiological surveillance of Zika, Chikungunya virus, and Dengue virus—suspected cases [13].

## Concurrent outbreaks of Dengue, Chikungunya and Zika virus infections—an unprecedented epidemic wave of mosquito-borne viruses in the pacific 2012—14

About 28 new documented outbreaks and circulation of Dengue, Chikungunya, and Zika virus infections have been reported between January 2012 and 17 September 2014 and about 120, 000 people were affected in the Pacific region. These outbreaks put extra burden on preexisting healthcare system in Pacific Islands. The risk for further spread in the Pacific Region is high due to several reasons. First could be due to low immunity as Dengue virus serotype 3 had been absent in this region since 1995. Secondly, in addition to *Aedes aegypti* and *Aedes albopictus* mainly in this region, local mosquitoes such as *Aedes polynesiensis* or *Aedes hensilli* can also transmit these viruses and thirdly, the large population mobilization and airline travel facilitate the spread [14].

## Dengue and Chikungunya viruses infections: long-distance spread and outbreaks in naïve areas

Outbreaks of Dengue and Chikungunya virus infections are taking place in previously disease-free areas. The important factor for long-distance spread of infectious disease is increased human mobility. Outbreaks were caused by infected persons coming from endemic and epidemic areas, acting as a Trojan horse for these germs. After the virus was incorporated in the new area, other factors like climate change, virus evolution, lack of vector control, socio-demographic changes, and environmental changes i.e., rapid uncontrolled urbanization, play an important role for geographic spread of mosquito-borne infections [15].

## Identifying and diagnosing the patient with unclear diagnosis

Due to long turnaround time, serum testing is not effective for emergent testing, therefore diagnosis of Dengue, Zika and Chikungunya virus infection should be made on clinical grounds. Even in strongly suspected cases of Zika and Chikungunya virus infections, Dengue virus infection still needs to be ruled out due to its life threatening complications. Acetaminophen should be preferred over nonsteroidal antiinflammatory drug given the hemorrhagic complications in Dengue virus infection cases. All three entities: Zika, Chikungunya, and Dengue virus infections are reportable diseases. If the patient also traveled to endemic for malaria, it is strongly recommended to rule out malaria. On the basis of travel, other illnesses like yellow fever, typhoid, leptospirosis, and helminth needs to be ruled out too [9].

Following testing algorithm is strongly recommended [9].

| Emergency department testing | Sending out testing |
|---|---|
| Malaria Smear/Point of care test | Dengue virus infection |
| Complete blood count (CBC) | Dengue NS1 polymerase chain reaction (PCR) |
| Chem 7 | Dengue Immunoglobulin M |
| Liver Function Tests (LFTs) | Chikungunya virus infection |
| Prothrombin time/Partial thromboplastin time (PTT) | Zika virus infection |
| Urine Analysis (UA)/human chorionic gonadotropin (Hcg) | |
| Chest radiograph | |

## Personal protection in endemic areas

Both as an issue to personal safety as well as public health measure, personal protection in an endemic area cannot be emphasized. The use of Environmental Protection Agency-registered insect repellant such as N,N-diethyl-meta-toluamide

(DEET) or picaridin are strongly recommended by Center for disease Control and Prevention. These agents are proven to be safe and effective for use in infants over two months, pregnant and breast feeding women [9]. A list of approved agents can be found at https://www.epa.gov/insect-repellents. Additional use of screens and mosquito netting are also important.

## Discovery of fifth serotype of Dengue virus (DENV-5): a new public health dilemma in Dengue virus infection control

Recently in October 2013, the fifth variant DENV-5 has been isolated which follow the sylvatic cycle contrary to other four serotypes, which follow the human cycle. Likely cause of emergence of these serotypes is due to genetic recombination, natural selection, and genetic bottlenecks. DENV-5 is currently present in India. Discovery of DENV-5 and more such sylvatic strains may impede the Dengue virus vaccine. Sustainable Dengue control then depends on integrated vector management [16].

## Mosquito-borne diseases and cancer: what do we really know?

There are only few studies published about relationship between some types of cancer and mosquito-borne diseases. The association between malaria and cancer may be explained by immune system suppression inducing by plasmodium. A second explanation by Lehrer [17,18] hypothesizes that Anopheles mosquitoes, a vector for malaria, could represent a source for brain tumor viruses. The evidence of association between Anopheles bites and brain tumors was reported from the link malaria outbreaks in USA and brain tumor incidence. Further research about this relationship is urgently needed as if the mosquito-transmitted brain tumor viruses are identified, development of brain tumor vaccine might be possible.

Espina et al. [19] observed increased production of tumor necrosis factor alpha in Dengue virus (serotype DEN-2)-infected human monocyte cultures. Monocytes play an important role in defense mechanisms against viruses by viral phagocytosis of infected apoptotic cells and release of proinflammatory cytokines. Later on, Chen et al. [20] described high viral DEN-2 titer, macrophage infiltration, and tumor necrosis factor alpha production in the local tissues as important events leading to hemorrhages [21].

## Dengue virus infection in the United States

Almost all Dengue virus infection reported in United States were acquired elsewhere by travelers and these imported cases rarely result in secondary transmission due to infrequent contact between Aedes and people [4].

## Dengue hemorrhagic fever — U.S.-Mexico border, 2005

The last reported Dengue virus infection outbreak was in south Texas in 2005. A case of Dengue hemorrhagic fever was reported in a resident of Brownsville, Texas in July 2005. On June 24, 2005, patient had acute onset of fever, chills, headache, nausea, vomiting, abdominal pain, arthralgia, and myalgia. In her youth, patient lived in across the border in the city of Matamoros in Tamaulipas, Mexico. Due to her illness, patient moved across the border into Matamoros. On June 28, she was hospitalized in Matamoros with the likely diagnosis of Dengue virus infection and urinary tract infection. Patient had thrombocytopenia (62,000 platelets/mm3) but no hemorrhagic manifestations, treated with antibiotics and discharged. On July 1, patient came back and sought treatment for continued fever, chills, vomiting, and abdominal pain and was hospitalized in Brownsville, Texas. Blood pressure was 94/70 mm Hg, and laboratory testing indicated proteinuria, hematuria, with a further decrease in platelet count (43,000/mm3). The patient was treated with antibiotics for a urinary tract infection and fluid resuscitation. Platelets dropped to 39,000/mm3 and albumin to 2.9 g/100 mL. A fecal occult blood test was positive, and pleural effusion was noted. Platelet count improved to 118,000/mm3. Patient was discharged on July 4 with diagnosis of possible murine typhus or viral infection and doxycycline prescription. Dengue virus infection was not diagnosed despite the patient's clinical characteristics (i.e., acute fever, platelet count <100,000/mm3, evidence of bleeding [hematuria and fecal occult blood] and plasma leakage) consistent with World Health Organization (WHO) criteria for Dengue hemorrhagic fever. Subsequently, serum sample taken on July 3 was positive for Dengue immunoglobulin M (IgM) by enzyme-linked immunosorbent assay (ELISA) and had an elevated titer of immunoglobulin G (IgG) antibodies to Dengue virus (1:655,350) and interpreted as indicative of a secondary Dengue virus infection. In August 2005, 1251 cases of Dengue fever, including 223 cases (17.8%) of Dengue hemorrhagic fever were reported in the neighboring state of Tamaulipas, Mexico [22].

A small outbreak of Dengue virus infection occurred in Hawaii in 2001. Dengue virus illness is endemic in Puerto Rico, the U.S. Virgin Islands, Samoa, and Guam. Large island-wide epidemics have been reported in Puerto Rico since late 1960s with most recent one in 2007 when more than 1000 cases were reported. The principal Dengue virus vector is *Ae. aegypti* in Puerto Rico and most of the Caribbean Basin. In Puerto Rico, Dengue virus infection transmission follows a seasonal pattern: low transmission from March till June and high from August till November [4].

## Texas lifestyle limits transmission of Dengue virus

Dengue virus infection like illness was first noted in the New World as major outbreak Philadelphia in 1780 R. In 1922, about 500,000 cases were noted in

Texas. From 1980 to 1999, only 64 locally acquired cases were reported in Texas whereas 62,514 suspected cases in three adjoining Mexican states—Coahuila, Nuevo León, and Tamaulipas [23].

Laredo, Texas, United States (population 200,000), and Nuevo Laredo, Tamaulipas, Mexico (population 289,000) are fundamentally a single city locally known as "los dos Laredos" divided by a small river, the Rio Grande. In the summer of 1999, a seroepidemiologic survey was conducted to study the factors affecting the Dengue virus infection transmission on either side of the border. IgM and IgG sero-positivity to Dengue virus was lower in Laredo than in Nuevo Laredo. Conversely, Texas side of the border has higher number of mosquito-infested containers. About 82% homes were air-conditioned in Laredo compared to 24% in Nuevo Laredo. Laredo had greater proportion of houses with intact screens, greater average distance between houses and a few persons residing in each house. The difference in transmission rate between the two cities less likely related to climate variation given their proximity. *Ae. aegypti* infestation rates in Laredo were markedly higher in despite the mosquito control programs. The likely reasons for lower transmission in Laredo are that most of the shops, houses, and restaurants were air-conditioned and close windows leading to less opportunity for mosquito/human contact. In Nuevo Laredo, most of the shops, bars, and restaurants are left open to street. Most of the buildings in United States are air-conditioned, fully glazed, and permanently closed. Even if the infected mosquitos gain entry to such buildings, they are less likely to survive in artificially dried atmosphere. The dollar cost of electricity is similar in Laredo and Nuevo Laredo but income per capita is much higher in Texas compared to Tamaulipas [23].

It has been stated earlier that as a result of global warming, Dengue virus infection, malaria and other mosquito-borne diseases will be more prevalent in United States [24–27]. These predictions are made from vectorial capacity model which population density, biting frequency, and daily survival probability of the vector, and extrinsic incubation period of the pathogen [28,29] but lacks in counting the effect of factors like air-conditioning, use of evaporative coolers, and behaviors of mosquitos and humans [30].

## Dengue surveillance in United States

Beginning 2009, all the nationally diagnosed Dengue virus infection cases will be reported to Centers for Disease Control and Prevention. Since 1915, statistical data is being compiled regarding cases in Puerto Ric and since 1969, Center for Disease Control and Prevention's Dengue Branch, located at San Juan and operates the island-wide passive Dengue viral illness surveillance system (PDSS) in collaboration with Puerto Rico Department of Health. Passive Dengue viral illness surveillance system played an important role in identifying the first case of Dengue virus infection in the Americas, the first cluster of cases of Dengue virus infection, first laboratory-confirmation of Dengue virus infection and Dengue related death in Puerto Rico [4].

## References

[1] Bäck AT, Lundkvist ÅKe. Dengue viruses—an overview. Infect Ecol Epidemiol 2013;3(1):19839.
[2] Organization, W.H.. Dengue. 2019.
[3] Yung C-F, et al. Dengue serotype-specific differences in clinical manifestation, laboratory parameters and risk of severe disease in adults, Singapore. Am J Trop Med 2015;92(5):999—1005.
[4] Control, C.f.D.. Dengue. 2019.
[5] Torrey T. Difference between an epidemic and a pandemic. 2019.
[6] Qureshi AI. Ebola virus disease. 2019.
[7] Kuno G. Review of the factors modulating dengue transmission. Epidemiol Rev 1995;17(2):321—35.
[8] Barthel A, et al. Breast milk as a possible route of vertical transmission of dengue virus? Clin Infect Dis 2013;57(3):415—7.
[9] Patterson J, Sammon M, Garg M. Dengue, zika and chikungunya: emerging arboviruses in the new world. West J Emerg Med 2016;17(6):671.
[10] Solomon T, Mallewa M. Dengue and other emerging flaviviruses. J Infect 2001;42(2):104—15.
[11] Singh S, Kumar AJV. Ocular manifestations of emerging flaviviruses and the blood-retinal barrier. Viruses 2018;10(10):530.
[12] Willcox AC, et al. Seroepidemiology of dengue, zika, and yellow fever viruses among children in the democratic republic of the Congo. Am J Trop Med Hyg 2018;99(3):756—63.
[13] Waggoner JJ, et al. Viremia and clinical presentation in nicaraguan patients infected with Zika virus, chikungunya virus, and dengue virus. Clin Infect Dis 2016;63(12):1584—90.
[14] Roth A, et al. Concurrent outbreaks of dengue, chikungunya and zika virus infections—an unprecedented epidemic wave of mosquito-borne viruses in the Pacific 2012—2014. Euro Surveill 2014;19(41):20929.
[15] Rezza GJP. Dengue and chikungunya: long-distance spread and outbreaks in naïve areas. Pathog Glob Health 2014;108(8):349—55.
[16] Mustafa MS, et al. Discovery of fifth serotype of dengue virus (DENV-5): a new public health dilemma in dengue control. Med J Armed Forces India 2015;71(1):67—70.
[17] Lehrer S. Anopheles mosquito transmission of brain tumor. Med Hypotheses 2010;74(1):167—8.
[18] Lehrer S. Association between malaria incidence and all cancer mortality in fifty U.S. States and the District of Columbia. Anticancer Res 2010;30(4):1371—3.
[19] Espina LM, et al. Increased apoptosis and expression of tumor necrosis factor-alpha caused by infection of cultured human monocytes with dengue virus. Am J Trop Med Hyg 2003;68(1):48—53.
[20] Chen HC, et al. Both virus and tumor necrosis factor alpha are critical for endothelium damage in a mouse model of dengue virus-induced hemorrhage. J Virol 2007;81(11):5518—26.
[21] Benelli G, et al. Mosquito vectors and the spread of cancer: an overlooked connection? Parasitol Res 2016;115(6):2131—7.

[22] Control, C.f.D. and Prevention. Dengue hemorrhagic fever–US-Mexico border, 2005. MMWR Morb Mortal Wkly Rep 2007;56(31):785.

[23] Reiter P, et al. Texas lifestyle limits transmission of dengue virus. Emerg Infect Dis 2003;9(1):86.

[24] Watson R, et al. Impacts, adaptations & mitigation of climate change: scientific-technical analyses, vol. 25 (2/3); 1998. p. 133.

[25] Jetten TH, Focks DA. Potential changes in the distribution of dengue transmission under climate warming. Am J Trop Med Hyg 1997;57(3):285−97.

[26] Patz JA, et al. Dengue fever epidemic potential as projected by general circulation models of global climate change. Environ Health Perspect 1998;106(3):147−53.

[27] Watson RT, et al. The regional impacts of climate change. 1998.

[28] Macdonald GJTE, Malaria Co. The epidemiology and control of malaria. 1957.

[29] Bailey NT. The mathematical theory of infectious diseases and its applications. 5a Crendon Street, High Wycombe: Charles Griffin & Company Ltd; 1975. Bucks HP13 6LE.

[30] Molineaux L. The pros and cons of modelling malaria transmission. Trans R Soc Trop Med Hyg 1985;79(6):743−7.

Chapter 4

# Global health-care perspective of Dengue viral disease

Sachin M. Bhagavan
*Resident Physician, Department of Neurology, University of Missouri Health Care, Columbia, MO, United States*

## Introduction

Dengue viral illness is one the most important arboviral diseases of the humans, occurring in tropical countries in more than 60 countries. An estimated 2.5 billion people worldwide are at risk of infection; approximately 975 million of them live in urban areas in tropical and subtropical countries in Southeast Asia, Africa, East Mediterranean, the Pacific, and the Americas. More than 50 million infections occur each year, including 500,000 hospitalizations for Dengue hemorrhagic fever, mainly among children, with the case fatality rate 10%–15% in endemic regions [1,2]. This chapter aims to highlight the impact of the Dengue viral illness worldwide and the factors that contribute to its global spread. This chapter also sheds some light on the measures that were taken in the past, reasons for their shortcoming, and at last on the research advances in the field of development of vaccines targeting Dengue viral illness.

## Global epidemiology and impact

Dengue viral illness has been a big concern globally. The disease burden was first estimated based on the constant infection rate annually on a crude population at risk like 10% in one billion persons, another estimate of 4% in 2 billion persons [3,4]. The estimated global infection rate in 1985 was estimated to be 80–100 million cases per year. More data being collected and sophisticated tools and methods are being used to identify the incidence of the disease in recent years. Knowing the geographical distribution and amount of disease burden is a key to understand global mortality and morbidity to efficiently allocate the resources available for Dengue viral disease management and in evaluating the impact of measures taken toward a goal of eliminating the disease.

**Dengue Virus Disease. https://doi.org/10.1016/B978-0-12-818270-3.00004-7**

In 2009, the World Health Organization established new guidelines for diagnosis and management, which changed the case definitions [1]. Though they are more comprehensive, these definitions had a few shortcomings, like altered hemostasis was not incorporated [2]. The actual numbers of viral illness cases are underreported because it takes into account only laboratory positive cases, whereas a fraction of Dengueviral illness cases present with just viral prodrome-like illness. For example, estimates made in 2013 mentioned that there are 390 million Dengue viral illness infections occurring every year (95% confidence interval 284−528 million), of which 24.6% manifest clinically with any severity of disease including Denguehemorrhagic fever/ Dengue viral illness shock syndrome which is 3 times more than estimated by the World Health Organization [5,6].

To estimate the global incidence, various approaches combining historical occurrences and expert prediction have been used to identify areas that are at maximum risk [7,8]. In the present time, various mathematical models have been used to estimate the global burden of the disease. These methodologies are expected to help in estimating the economic impact of the disease, assessing morbidity and mortality, estimating amount of vaccines required, data for research and vaccine development.

Fig. 4.1 compares the global incidence of Dengue viral illness in 2012 and 2017 which has been developed from a model from the Global Health Data 2017. This model used incidence from officially reported cases and adjusted raw estimates for underreporting based on published estimates of expansion factors. To correct the underreporting, investigators used a three-phase approach. First, expected spatial distribution of the disease was defined based on a principal component analysis of population-weighted probability of Dengue viral illness transmission. Secondly, the association between this expected distribution and reported incidence was modeled using mixed-effects binomial modes; the model was calibrated by benchmarking those deviations against published empirical expansion factors. Lastly, the model distributed total cases to cases based on age−sex groups by using the age−sex distribution of Dengue viral illness cases captured by Hospital Information System of the Brazilian Unified National Health System [9].

Currently the incidence of the disease globally has been on the rise. According to Global Health Data 2017 as shown in Fig. 4.2, the incidence per 100,000 populations was less than 500 in 1990, which has risen in the last 25 years to 1328 in 2017. This corresponds to more than 104 million new cases seen worldwide in just the year 2017 [10].

As we can see from Fig. 4.3 and Table 4.1, the incidence has been on the rise globally in the last 25 years in all the major regions of the world. Dengue viral illness has had significant impact in South Asia, Latin America, Africa, and the Middle East. South and Southeast Asian countries reported the highest incidence rate in the years with as high as 3670 per 100,000 populations as compared with 1009 per 100,000 populations in the Americas. There is a

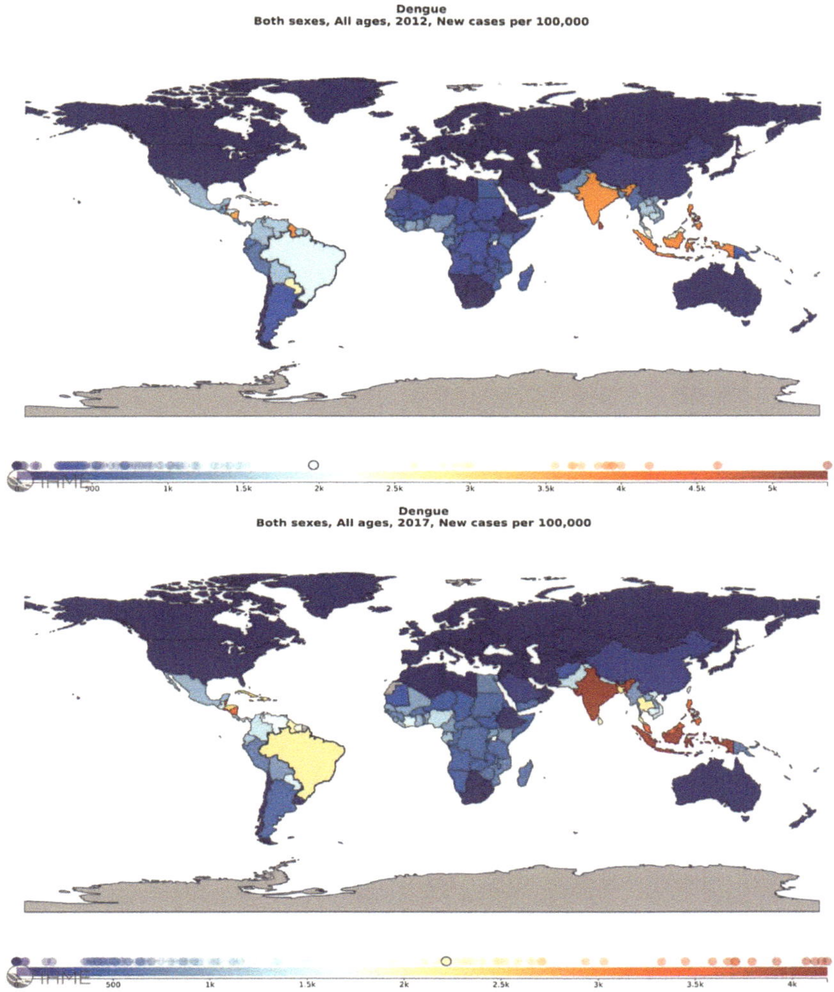


FIGURE 4.1 Comparison of global incidence of Dengue viral disease in 2012 (top figure) and 2017 (bottom figure).

significant increase of reporting in the number of cases from India and the eastern Mediterranean region due to numerous outbreaks, with recent reports of cases in Sudan, Yemen, Madagascar, Pakistan, and Saudi Arabia. Although Africa reported an incidence of 688 per 100,000 in 2017, the lack of appropriate diagnostic test and adequate clinical suspicion still underestimated the true burden of the disease [1]. There has been prospective cohort studies in Thailand and Nicaragua that show an incidence of Dengue virus infection of 6%–29% per year [11,12]. There has been a resurgence of Dengue viral illness

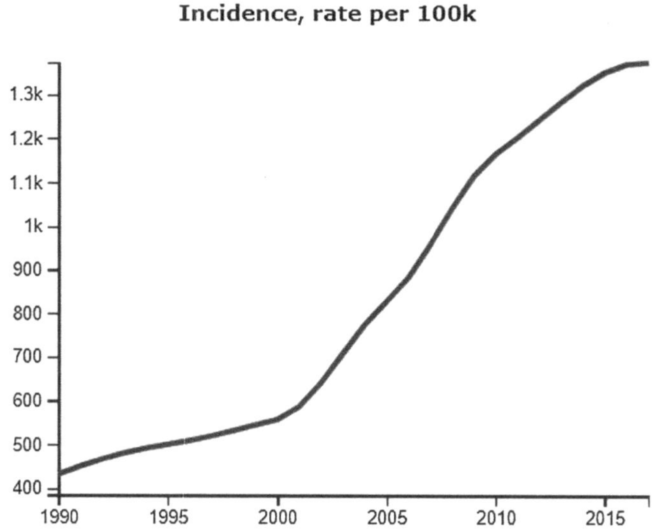


FIGURE 4.2 Incidence per 100,000 population globally.

activity documented in Buenos Aires, Hong Kong, the Galápagos Islands, Easter Island, and Hawaii. Dengue viral illness has also been reported in Florida, southeastern France, and Madeira Island [1,13]. Increasingly, coinfections of Dengue virus infection have been occurring with malaria, human immunodeficiency virus/acquired immunodeficiency syndrome, leptospirosis, and Chikungunya., Dengue viral transmission by blood transfusion is also reported.[14].

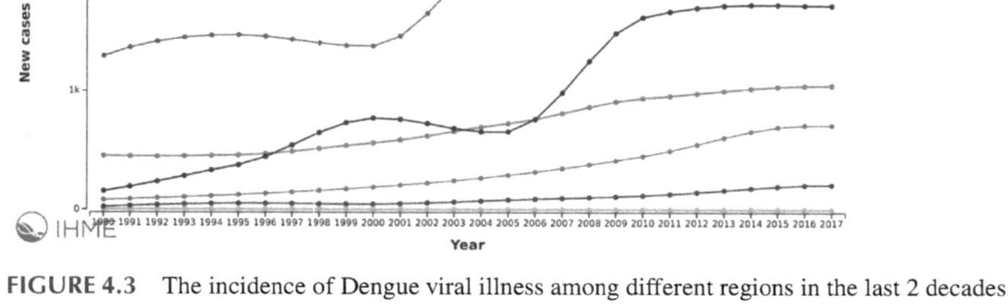


FIGURE 4.3 The incidence of Dengue viral illness among different regions in the last 2 decades.

**TABLE 4.1** The incidence of Dengue viral illness in different World Health Organization regions in 2017.

| Region | Incidence per 100,000 in 2017 |
|---|---|
| African region | 688.74 |
| Region of the Americas | 1009.87 |
| Southeast Asia region | 3670.69 |
| Eastern Mediterranean region | 596.21 |
| Western Pacific region | 483.65 |

Along with the increase in incidence, mortality has also been increasing in all regions of the world. According to Global Health Data 2017, Figs. 4.4 and 4.5 show global mortality and region-specific mortality caused by Dengue viral illness. This model has used Cause of Death database and the Cause of Death Ensemble Modeling tool [15,16]. Covariates on basis of expected associations with Dengue viral illness mortality and biological plausibility were included like environmental covariates (rainfall, latitude, people in urban areas, etc) and country level of development (income per capita, mean education level, access to health system). Population weighted mean probability of Dengue viral illness transmission was included. All cause mortality was

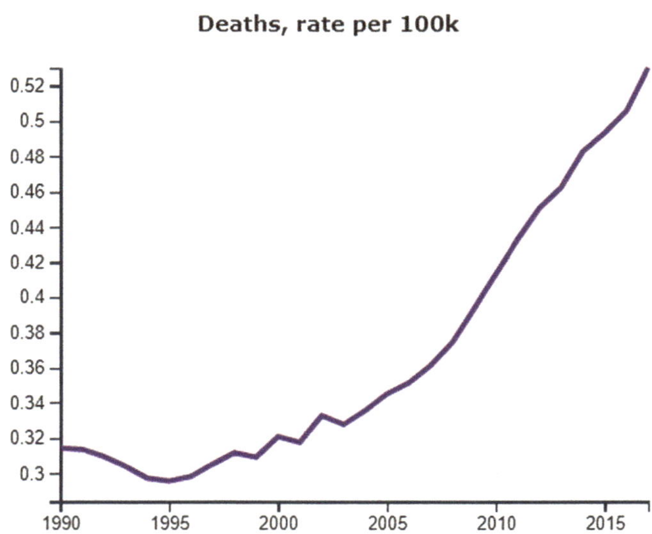


FIGURE 4.4 Trend in deaths due to Dengue viral illness globally.

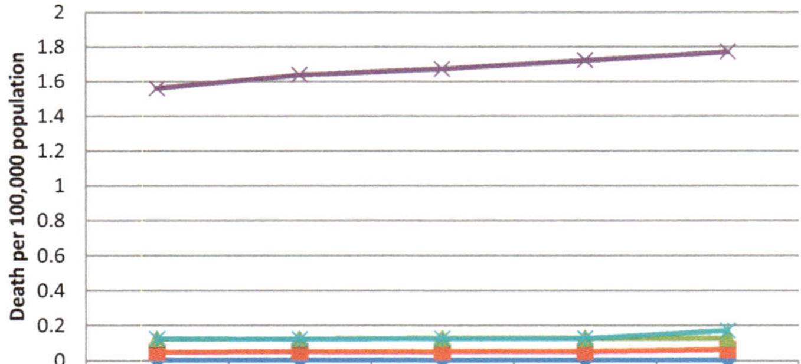


| | 2013 | 2014 | 2015 | 2016 | 2017 |
|---|---|---|---|---|---|
| African Region | 0.005103633 | 0.005127765 | 0.005136629 | 0.005087465 | 0.005165073 |
| European Region | 0.048761881 | 0.050273023 | 0.050971208 | 0.052334086 | 0.062462349 |
| Region of the Americas | 0.124617261 | 0.124242324 | 0.128473305 | 0.127975749 | 0.127488048 |
| South-East Asia Region | 1.559522856 | 1.637419794 | 1.671872119 | 1.720608131 | 1.768996711 |
| Western Pacific Region | 0.125692711 | 0.124373739 | 0.127398976 | 0.126439226 | 0.172438814 |

**FIGURE 4.5** Region-specific Dengue viral illness mortality rate in the last 5 years.

estimated and then within each sex-age-country-year group, the total of all-cause-specific death was constrained to equal the number of all-cause death through a process CodCorrect [15].

Mortality has been rising exponentially in the last 10 years with mortality rate as high as 0.53 per 100,000 population in 2017 as compared to 0.34 per 100,000 in 2005. As correlated with incidence, mortality was highest in those countries in South East Asia with as high as 1.76 per 100,000 populations as compared to other regions where mortality was in the range of 0.05−0.017 (as shown in Fig. 4.4). Among the Southeast Asian countries, Indonesia, Malaysia, Philippines, Bali, and India were the most affected with mortality rate as high as 9 per 100,000 in certain regions.

When we look into the morbidity of the disease, measured in terms of Daily Adjusted Life Years, it has also been increasing in the last 10−15 years. According to Global Health Data 2017, Daily Adjusted Life Years in 2017 was 38.25 per 100,000 population as compared to 29.02 per 100,000 in the year 2005 as seen in Fig. 4.6. This model calculated Daily Adjusted Life Years as the sum of Years to Live and Years Lived with Disability. Years to Live was estimated as the difference between age at death and life expectancy for those surviving to that age at death [15]. Years Lived with Disability was estimated by assigning a health state and corresponding disability weight to each case, which was based on pooled results from Global Burden of Disease 2010 Disability Weights Measurement Study [17] and more recent European Disability Weights Study [18]. Each case was assigned two acute health states: moderate and severe episodes [19]. Proportion for the split between moderate

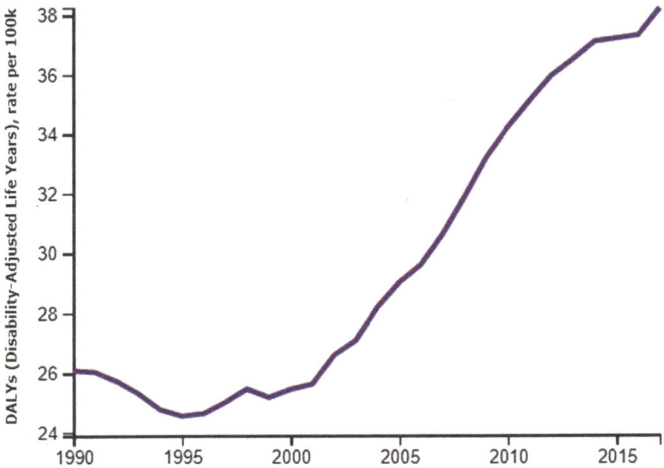


FIGURE 4.6 Trend in Disability Adjusted Life Years globally.

and severe was derived from a meta-analysis of the subset of data that had both total number of cases and number of severe cases. Lastly, for postDengue viral illness chronic fatigue was assigned the disability weight for infectious diseaseand post acute consequences.

Among the different regions of the world, Southeast Asia had the highest effect with 114.25 per 100,000 population. Daily Adjusted Life Years in different regions of the world has been shown in Table 4.2.

The disease burden mentioned above takes into account only the acute illness of 7 days or less asstandard. There are increasing reports of long-term

**TABLE 4.2 Daily Adjusted Life Years in various regions in 2017.**

| Regions | Daily adjusted life years in 2017 per 100,000 population |
|---|---|
| European region | 0.092188 |
| Southeast Asia region | 114.8533165 |
| Eastern Mediterranean region | 9.216687285 |
| Western Pacific region | 16.95258218 |
| African region | 7.122431707 |
| Region of the Americas | 15.57597189 |

postviral effects, such as depression and chronic fatigue syndrome, which add up further to the impact on productivity [20]. Also, disease burden does not account for lost work, absence from school, or lost tourism. The overall estimated disease burden is comparable to major diseases like tuberculosis or malaria, although those diseases receive major funding worldwide [21].

Numerous reasons can be attributed for this increase in incidence, morbidity, and mortality of Dengue viral illness. As we know, Dengue virus is present in four serotypes, DEN 1–4. Each of the four serotypes has been evolved into multiple genotypes, which is one of the major factors contributing to the global burden of the disease. Infection with one strain of virus does provide subsequent immunity to that particular strain, but does not provide adequate immunity to other strains. In fact, it causes "sensitization" where subsequent infections with a different strain creates a more serious and critical impact in form of "Dengue hemorrhagic fever/Dengue shock syndrome" contributing to the morbidity and mortality. Dengue viral illness epidemics occur in those countries where clinical management of Dengue hemorrhagic fever is at times suboptimal that often leads to increased mortality. Partly because Dengue hemorrhagic fever has a very nonspecific constellation of symptoms developing rapidly leading to shock and ultimately death within few hours. Global warming has also aided in the wider geographic distribution of Aedes mosquitoes, thereby increasing Dengue viral illness epidemic potential in temperate regions. Increased globalization, high population density, rural to urban migration, and development of rapid transport systems and urbanization have also led to global spreading of the disease. Traveling has made a significant impact in the spread of the disease, with travelers harboring the virus going to nonendemic areas constituting the main source for triggering transmission. International travelers are most at risk of Dengue viral illness, with attack rates reported as high as 5·51 cases per 1000 travel-months. Dengue viral illness has now become the leading cause of fever in returning travelers, having overtaken malaria for travelers to Southeast Asia.

## Global economic impact

Dengue viral illness has also made a huge impact globally in terms of economy. Although it is difficult to evaluate the true economic burden of the disease on each nation, efforts have been made in the past to estimate the economic aspect of the disease. Limited availability, inadequate resources for data collection in many countries, and poor availability of specific guidelines and protocols in diagnosing the disease are the challenges faced by the nations that hinder estimating the amount of money invested for the disease. In addition to the cost borne by the health-care system, there is family economic loss to be considered. One study estimated 14.8 days lost in ambulatory cases and 18.9 days lost in hospitalized cases. The impact on the family by one Dengue viral illness episode can be estimated as high as three times the family monthly income. The disease leads to families selling their belongings/taking out a loan for repayment as 45% of health costs are borne by patients or their family globally. Among those who do not seek health care, cost remains hidden. Most economic studies estimate the disease burden by looking at anywhere from 20 to 70 countries while Dengue viral illness transmission is present in almost 140 nations.

Therefore studies were done where extrapolation of available data and estimate through various mathematical models help in estimating and analyzing the disease somewhat accurately. One such study was done by Shepard et al. [22] where the study used various factors to estimate the economic burden. It estimated the incidence of Dengue viral illness by taking into account the reported episodes by the World Health Organization, Ministry of National Health Statistics and published literature, and estimated the underreporting through various literature and formulated aDengue viral illness score which comprised the probability of Dengue viral illness in an area, its population density, and death estimated. The study estimated that global average cost per Dengue viral illness case is $333 (95% confidence interval 283–403) for cases admitted to hospital, while for ambulatory cases it is $60 (95% confidence interval 54–68). In case of death by Dengue viral illness, long-term cost was estimated to be $80,414 for death of a child while death of an adult was estimated to be $75, 820 [22]. These figures show that the financial burden of the Dengue viral illness ($8.9 billion) [22] is more than other major infectious disease such as rotavirus gastroenteritis ($2billion) [23], cholera ($3.1 billion) [24], rabies ($8.6 billion) [25], and Chagas disease (7.2 billion) [26] (Fig. 4.7, Table 4.3).

## Initial global response

Due to the concerns of the burden of the Dengue viral illness virus, historically a unique mosquito control program was initiated in the 1940s. By 1960, efforts of the Pan American Health Organization, *Aedes aegypti* had been eradicated from major South and Central American countries. Many programs were reduced after this achievement and their goal was directed to a different problem. Within 2 decades, the vector regained its former population.

The World Health Organization in 2012 came up with a global strategy primarily to reduce Dengue viral illness–related mortality by 50% and morbidity by 25% by 2020 [6]. Primary focus was on preventive measures, early warning systems, and risk assessment in the form of early case detection and appropriate identification and management of severe cases. Secondary focus was on capacity building, investment in health-care infrastructure, and research for better understanding and combating Dengue viral illness. The plan also called for estimating the true burden of the disease, which would help to know the extent of the disease and to see if the policy is effective in reducing the burden. Therefore it emphasizes the importance of locoregional

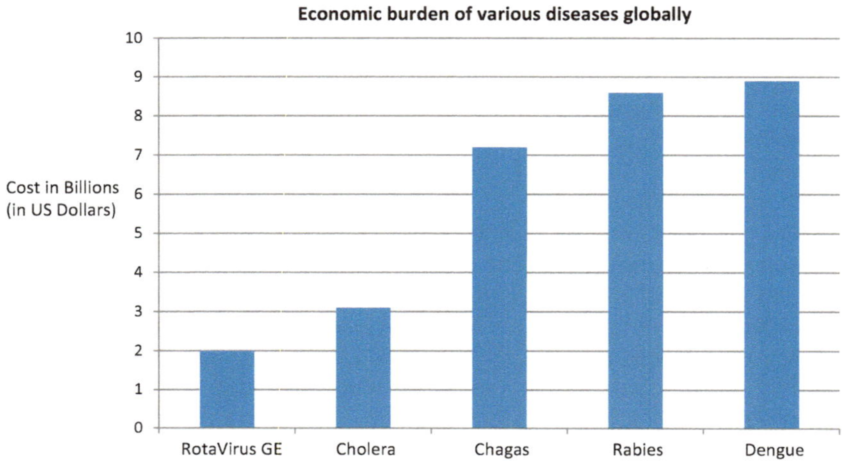


FIGURE 4.7 The cost of Dengue viral illness compared to other diseases globally.

involvement where the ministries of health in endemic countries coordinate with the World Health Organization for more accurate estimate of the disease and implementation of the strategies at the level of the public.

In recent years, several new tools and strategies for Dengue viral illness control and prevention have been developed and are available to public health practitioners and clinicians. One of the important highlights is the development of rapid commercial diagnostic tests, which is used in many endemic countries. These tests have become very useful as studies in various countries show that the test is 60%–70% sensitive and >95% specific. The World Health Organization has included dengue viral illness in its "pocket book of hospital care", which health-care workers in endemic countries have been using for the management of dengue viral illness fever [27,28]. An audiovisual guide and transcript for health-care workers responding to outbreaks has been an effective tool in early recognition, diagnosis, and hospital management [28].

TABLE 4.3 The average cost of Dengue viral illness globally.

| Global average cost | Value (in US dollars) |
|---|---|
| Admission to hospital | 333 (CI 283–403) |
| Ambulatory cases | 60 (CI 54–68) |
| Death of child | 80,414 |
| Death of adult | 75,820 |

The World Health Organization created "Global Outbreak Alert and Response Network", which aims at responding to a developing outbreak very swiftly to prevent the outbreak itself or limit the consequences in case of an outbreak. The network developed an "integrated vector management approach" in 2004 where the main aim was to reduce or interrupt transmission of the disease. The methodology emphasized on eliminating breeding grounds, use of indoor residual sprays, and also use of an obligate intracellular bacterium *Wolbachia* which when injected intracellular into male Aedes mosquito is thought to bring a reproductive manipulation rendering offspring nonviable thereby reducing the mosquito burden [29,30]. .This method seems to have a beneficial impact on the rate of Dengue viral illness as evident in Brazil, who released its first Wolbachia mosquito in 2014. Similar efforts have been seen in Indonesia under the "Eliminate Dengue viral illness Program". Singapore conducts a nation-based survey to identify the items around the house that are the most proficient breeding sites such as flowerpots and ornamental containers.

The World Health Organization has stressed the importance of research in the field of dengue viral illness. Historically, Dengue viral illness research has always been underfunded due to underestimating of the disease burden and its impact on society. However, more and more research projects have been undertaken in the last few years. The US National Institute of Allergy and Infectious disease has funded over 60 projects with emphasis on tetravalent vaccine development. Preventing outbreak, early detection, and to improve vaccine delivery remains a priority for the Bill and Melinda Gates Foundation. The European Commission provided 18 million Euros toward "Comprehensive control of Dengue viral illness Fever under changing climactic condition". With the aim to generate awareness and improve research funding, the Association of Southeast Asian Nations has designated June 15 to be "Dengue viral illness day" [31].

Finally, most effective method to combat any disease is to generate awareness among people and Dengue viral illness is not an exception. Another reason for the rising epidemics is the difficulty to control mosquito breeding. Though government from endemic nations have invested a lot of money for insecticide spraying, common people do not have the awareness of preventing breeding grounds for mosquitoes that perpetuate their cycles. Efforts for awareness have been attempted on an international platform during the 2014 World cup in Brazil and in the 2016 Summer Olympics. Social media can be used not only as a platform to increase awareness but also to report cases that could serve as a warning sign for potential outbreaks. The World Health Organization has emphasized a lot on awareness and promoted community-based, "bottom-up" communication for behavioral impact [32]. It primarily encourages active participation of community members in public health messaging. For example, in Thailand, there is one village health worker for every 10 households with the responsibility toward generating awareness and warning about outbreaks [32].

## Global innovations/interventions

Numerous efforts are being taken to prevent disease transmission, thereby reducing the impact on human lives. For adequate control over the disease, early identification of the epidemic is the first step. However, majority of the detection of the cases nationally rely on hospital-based reporting. Ineffective communication, untimely reporting, and frequent post hoc revisions limit identification and optimization of necessary interventions.

Therefore, an ideal tool should have the following characteristics: provide accurate data to regional/national level, ability to detect outbreaks and swift warning, be updated in real time, and reduce resource related delays.

Many epidemiological methods have been attempted to act as a supplemental tool by reducing the limitations of the traditional system. Autoregressive models like Seasonal Autoregressive Integrated Moving Average take into account seasonal patterns and help in determining useful incidence estimates [33–36]. Numerous and varied mechanistic models have been explored [37] and some long-term weather-driven models such as El Niño with Dengue viral levels in various countries [38].

Real-time Internet searches for Dengue viral illness tracking has been developing as an effective tool for disease identification and warning. Internet search is efficient, consistent, and gives a snapshot of real-time trends, thereby posing as a very strong supplemental tool. Studies have been done previously for evaluating the role of using Internet search data to track Dengue viral diseases [39,40]. Google Dengue Trends was started in 2011 and was one of the first tools to be used in quantification of Dengue viral illness in multiple regions of the world, thereby allowing a large section of population to access the data globally. Currently, live status and the trend of the disease are easily available which help in early identification and taking appropriate measures to reduce the risk of the disease outbreak. One such method can be seen in Fig. 4.8.

It is imperative to evaluate the combination of traditional methods and real-time Internet searches for detection of cases thereby combining the respective strengths of the data source. One such attempt is made to combine the autoregressive models with real-time Google search queries to explore the effectiveness of the combination [41]. The results are promising, but yet many facets have to be explored in order to create an efficient way of rapid, early detection with fast communication to the general population at risk to minimize risk and reduce burden of the disease.

## Global research and vaccine development

Dengue viral illness has emerged as a global public health problem in the last 20 years. This has been caused by the expanding geographic distribution of both the viruses and the principal mosquito vectors subsequent to global

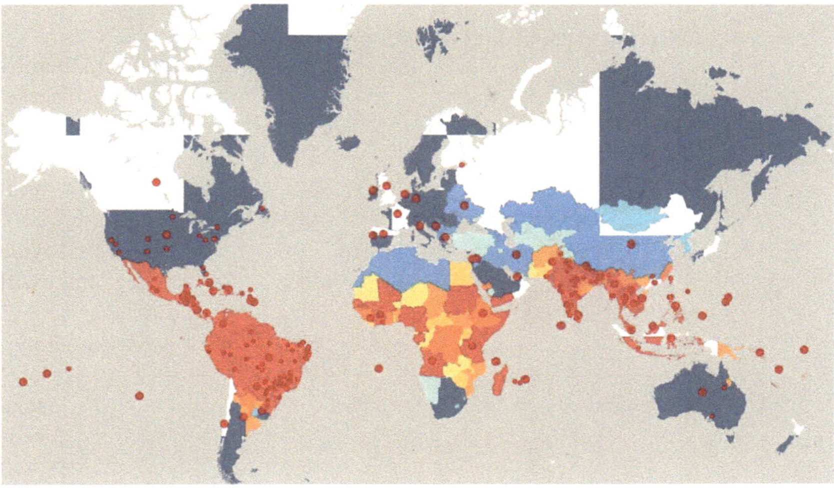

FIGURE 4.8 Trend of the Dengue viral illness in the last 3 months. *Source: https://www.healthmap.org/dengue/en/.*

demographic and societal changes. During this period, most tropical urban centers of the world have become hyperendemic, thus increasing the risk of epidemic transmission and the emergence of Dengue hemorrhagic fever. It is important to know the global impact and economic burden because such estimates are needed by policy planners to help allocate limited resources for research, prevention, and control activities [42].

The Scientific Working Group was organized by the United Nations International Children Emergency Funds/United Nation Development Program/World Bank/World Health Organization, a Special Program for Research and Training in Tropical Diseases in Geneva. The priority of dengue viral illness research areas are organized along four major research streams which will provide evidence and information for policy-makers and control programmes leading to more cost-effective strategies which will reverse the epidemiological trend [27,28,43–46]. These streams are highlighted as shown in Table 4.4.

As a result of the failure of vector control, the continuing spread and increasing intensity of Dengue viral illness has renewed interest and investment in Dengue viral illness vaccine development, making a safe, effective, and affordable tetravalent dengue viral illness vaccine a global public health priority. Dengue viral illness vaccine development has been in progress for several decades; however, the complex pathology of the illness, the need to control four virus serotypes simultaneously, and insufficient investment by vaccine developers have hampered progress. The available data suggest that neutralizing antibodies are the major contributors to protective immunity; however, the role of the cellular immune response requires further study. In this context, clinical

**TABLE 4.4 Different streams and activity performed.**

| Streams | Activities performed | Related information |
|---|---|---|
| Stream 1 | *Research related to reducing disease severity and case fatality* | *Optimization of clinical management* |
| Stream 2 | *Research related to transmission control through improved vector management* | *Development and evaluation of vector control tools and strategies* |
| Stream 3 | *Research related to primary and secondary prevention* | *Vaccines* |
| Stream 4 | *Health policy research contributing to adequate public health response* | *Health policies* |

trials are crucial for vaccine development owing to the unique information they provide on immune responses and reactogenicity. Also, long-term observations of vaccinated populations will be required to demonstrate the absence of antibody dependent enhancement (ADE) or severe disease [47−55].

An ideal vaccine for Dengue viral illness virus must be to tetravalent because each of these serotypes is present throughout the world and each can cause disease. The vaccine should include neutralizing antibody levels compatible with those observed in wild-type virus infection to limit the risk of adverse drug effect. Finally the vaccine should be available at a low cost for the developing world. Multiple live attenuated vaccine candidates are presently being evaluated in current clinical trials. Clinical safety and strong immunogenicity have been observed for empirically derived vaccine stringent for recombinant viruses using either genetically modified full-length vaccine strains or antigenic chimeric viruses. Inactivated monovalent vaccines and recombinant subunit vaccine consisting of purified and blood proteins are scheduled to be tested in clinical trial soon. Other approaches are being evaluated pretty clinically. Considerable progress has been made in the development of Dengue viral illness vaccine candidates in the last few years. One or more of the strategies being currently perceived should be successful in reducing the disease burden caused by this emerging pathogen [56,57].

Vaccine candidates should be evaluated in population-based efficacy trials in several at-risk populations in different geographical settings, including Asia and the Americas, which experience different patterns of Dengue viral illness transmission intensity and Dengue viral illness virus circulation. Vaccine developers are working with the pediatric Dengue viral illness Vaccine Initiative to establish suitable field sites. Developers are also working with the World Health Organization/Initiative for Vaccine Research to define the immunological correlates for protection and clinical trial design. Because of the important role of neutralizing antibodies as surrogates of protection, the validation of neutralization tests is a priority. Current approaches to vaccine development involve using live attenuated viruses, inactivated viruses, subunit vaccines, deoxyribonucleic acid vaccines, cloned engineered viruses, and chimeric viruses using yellow fever vaccine and attenuated Dengue viral illness viruses as backbones. Currently vaccines in different stages of clinical trials are shown in Table 4.5 [19,30,57–76].

The United States Food and Drug Administration approved Dengvaxia, the first vaccine to be approved for the prevention of Dengue viral illness disease caused by all the dengue viral illness serotypes (1, 2, 3, and 4) in people ages 9 through 16 years who have laboratory confirmed previous Dengue viral illness infection and who live in endemic areas. Dengvaxia is a live, attenuated vaccine that is administered as three separate injections, with the initial dose followed by two additional shots given 6 and 12 months later. The safety and effectiveness of the vaccine was determined in three randomized, placebo-controlled studies involving approximately 35,000 individuals living in endemic areas. The vaccine was determined to be approximately 76% effective in preventing symptomatic laboratory-confirmed Dengue viral illness disease in 9–16 years of age. Dengue vaccine has already been approved in 19 countries

**TABLE 4.5 Dengue viral illness vaccines in research.**

| | |
|---|---|
| Preclinical | Subunit recombinant antigen (domain III) vaccine by IPK/CIGB |
| | Live attenuated chimeric yellow fever/Dengue virus vaccine (YF-DEN) by Oswald Cruz foundation |
| | Tetravalent DNA vaccine by United States Naval Medical Research Center (US NRMC) & GenPhar |
| | Purified inactivated tetravalent vaccine by Walter Reed Army Institute of Research (WRAIR) and GlaxoSmithKline (GSK). |
| PHASE I | Live attenuated tetravalent vaccine comprising 3′ deletion mutations and DEN-DEN chimeras by United States National Institute of Health Laboratory of Infectious Disease (US NIH LID) and National Institute of Allergy and Infectious Diseases (NIAID) |
| | Live attenuated chimeric DEN2-DEN vaccine by CDC & Inviragen |
| | Recombinant E subunit vaccine by Merck |
| PHASE II | Live attenuated tetravalent chimeric yellow fever/Dengue virus vaccine (YF-DEN) by Sanofi Pasteur |
| | Live attenuated tetravalent viral isolate vaccine byWalter Reed Army Institute of Research (WRAIR) and GlaxoSmithKline (GSK). |

and the European Union. The most commonly reported side effects by those who received Dengvaxia are headache, muscle pain, joint pain, fatigue, injection site pain, and low-grade fever. The frequency of side effects was similar across different populations who received Dengvaxia and these side effects intended to decrease after each subsequent dose of the vaccine [77].

Dengvaxia is not approved for use in individuals who are not previously infected with Dengue viral illness virus as it appears to act like a first-degree infection without actually infecting the person with wild-type by dengue viral illness virus such that a subsequent infection can result in severe degree of disease. Therefore, health-care professional should evaluate individuals for prior infections to avoid vaccinating individuals who have not been previously infected by Dengue virus. This can be accessed through medical records for previous laboratory confirmed negative infection or through serological testing (tests using blood samples from the patient) prior to vaccination [77].

## Conclusion

Dengue viral illness is now endemic in more than 125 countries globally. Reasons for the currently observed and predicted expansion are multifactorial. These reasons may include climate change, virus evolution, and societal factors such as rapid urbanization, population growth and development, socioeconomic factors, as well as global travel and trade. The known social, economic, and disease burden of Dengue viral illness internationally is alarming and it is evident that the wider impact of this disease is grossly underestimated. Dengue viral illness has been like a silent but potent weapon impacting lives worldwide, with incidence rates climbing in the last decade. Multiple attempts and numerous measures to eliminate the disease have been unsuccessful. The World Health Organization Global Strategy for Dengue viral illness Prevention and Control, 2012–20 highlights the need for improved estimates of the true burden of Dengue viral illness globally due to the currently presumed underrepresentation. Surveillance and reporting is paramount for effective Dengue viral illness control, and more accurate quantification of the impact of Dengue viral illness globally will allow improved political, financial, and research prioritization as well as informed decision-making and enhanced modeling. Active participation of nations globally along with political and economical resources will be a key to eradicate this disease in future [27,28,48,56].

## References

[1] Dengue: guidelines for diagnosis, treatment, prevention and control. New Edition. Geneva: World Health Organization; 2009.

[2] Halstead SB, Cohen SN. Dengue hemorrhagic fever at 60 years: early evolution of concepts of causation and treatment. Microbiol Mol Biol Rev 2015;79:281–91.

[3] Halstead SB. Pathogenesis of dengue: challenges to molecular biology. Science 1988;239:476–81 [PubMed: 3277268].
[4] Monath TP. Yellow fever and dengue-the interactions of virus, vector and host in the re-emergence of epidemic disease. Sem Virol 1994;5:133–45.
[5] Bhatt S, Gething PW, Brady OJ, Messina JP, Farlow AW, Moyes CL etal The global distribution and burden of dengue. Nature; 496:504-507.
[6] World Health Organization. Global strategy for dengue prevention and control, 2012–2020. Geneva: WHO Press; 2012.
[7] Beatty ME, Letson GW, Margolis HS. Estimating the global burden of dengue. Am J Trop Med Hyg 2009;81(5 Suppl. 1):231.
[8] W.H.O. International travel and health: situation as on 1 January 2012. World Health Organization; 2012.
[9] The global burden of dengue: an analysis from the global burden of disease study 2013. Lancet Infect Dis. Stanaway JD, Shepard DS, Undurraga EA, Halasa YA, Coffeng LE, Brady OJ, Hay SI, Bedi N, Bensenor IM, Castañeda-Orjuela CA, Chuang TW, Gibney KB, Memish ZA, Rafay A, Ukwaja KN, Yonemoto N, Murray CJL.
[10] Global Health Data http://ghdx.healthdata.org.
[11] Endy TP, Anderson KB, Nisalak A, et al. Determinants of inapparent and symptomatic dengue infection in a prospective study of primary school children in Kamphaeng Phet, Thailand. PLoS Negl Trop Dis 2011;5:e975.
[12] Balmaseda A, Standish K, Mercado JC, et al. Trends in patterns of dengue transmission over 4 years in a pediatric cohort study in Nicaragua. J Infect Dis 2010;201:5–14.
[13] Alves MJ, Fernandes PL, Amaro F, et al. Clinical presentation and laboratory findings for the first autochthonous cases of dengue fever in Madeira island, Portugal, October 2012. Euro Surveill 2013;18:6.
[14] Lanteri MC, Busch MP. Dengue in the context of "safe blood" and global epidemiology: to screen or not to screen? Transfusion 2012;52:1634–9.
[15] GBD 2013 Mortality and Causes of Death Collaborators. Global, regional, and national age–sex specific all-cause and cause-specific mortality for 240 causes of death, 1990–2013: a systematic analysis for the Global Burden of Disease Study 2013. Lancet 2015;385:117–71.
[16] Foreman KJ, Lozano R, Lopez AD, Murray CJ. Modeling causes of death: an integrated approach using CODEm. Popul Health Metr 2012;10:1.
[17] Salomon JA. New disability weights for the global burden of disease. Bull World Health Organ 2010;88:879.
[18] Haagsma JA, de Noordhout CM, Polinder S, et al. Assessing disability weights based on the responses of 30,660 people from four European countries. Popul Health Metr 2015;13:10.
[19] Whitehead SS, Blaney JE, Durbin AP, Murphy BR. Prospects for a dengue virus vaccine. Nat Rev Microbiol 2007;5:518–28.
[20] World Health Organization. Marking ASEAN dengue day. 2013. https://www.who.int/westernpacific/news/detail/14-06-2013-marking-asean-dengue-day.
[21] Murray NE, Quam MB, Wilder-Smith A. Epidemiology of Dengue: past, present and future prospects. Clin Epidemiol 2013;5:299–309.
[22] Shepard DS1, Undurraga EA2, Halasa YA2, Stanaway JD3. The global economic burden of dengue: a systematic analysis. Lancet Infect Dis August 2016;16(8):935–41. https://doi.org/10.1016/S1473-3099(16)00146-8. Epub 2016 Apr 16.
[23] Grimwood K, Lambert SB, Milne RJ. Rotavirus infections and vaccines. Pediatr Drugs 2010;12:235–56.

[24] UPMC Center for Health Security. Infectious disease cost calculator. 2015. http://www.idcostcalc.org/contents/about/cost-of-ID.html.

[25] Hampson K, Coudeville L, Lembo T, et al. Estimating the global burden of endemic canine rabies. PLoS Negl Trop Dis 2015;9:e0003709.

[26] Lee BY, Bacon KM, Bottazzi ME, Hotez PJ. Global economic burden of Chagas disease: a computational simulation model. Lancet Infect Dis 2013;13:342–8.

[27] WHO. Dengue. In: Pocket book of hospital care for children. Guidelines for the management of common illnesses with limited resources. Geneva: World Health Organization; 2006a. p. 166–71.

[28] WHO. Dengue haemorrhagic fever: early recognition, diagnosis and hospital management. 2006. An audiovisual guide for health care workers responding to outbreaks. WHO/CDS/EPR/2006.4.

[29] Johnson KN. The impact of Wolbachia on virus infection in mosquitoes. Viruses 2015;7:5705–17.

[30] WHO. Global strategic framework for integrated vector management. Geneva, Switzerland: WHO; 2004.

[31] Warkentien T, Pavlicek R. Dengue fever: historical perspective and the global response. J Infect Dis Epidemiol 2016;2:015.

[32] WHO Regional Office for South-East Asia. Comprehensive guidelines for prevention and control of dengue and dengue haemorrhagic fever revised and expanded edition. 2011.

[33] Johansson MA, Reich NG, Hota A, Brownstein JS, Santillana M. Evaluating the performance of infectious disease forecasts: a comparison of climate-driven and seasonal dengue forecasts for Mexico. Sci Rep 2016;6. https://doi.org/10.1038/srep33707.

[34] Promprou S, Jaroensutasinee M, Jaroensutasinee K. Forecasting dengue haemorrhagic fever cases in Southern Thailand using ARIMA Models. Dengue Bull 2006;30:99.

[35] Luz PM, Mendes BVM, Codeço CT, Struchiner CJ, Galvani AP. Time series analysis of dengue incidence in Rio de Janeiro, Brazil. Am J Trop Med Hyg 2008;79(6):933–9. PMID: 19052308 18. Choudhury ZM, Banu S, Islam AM. Forecasting dengue incidence in Dhaka, Bangladesh: A time series analysis. 2008.

[36] Eastin MD, Delmelle E, Casas I, Wexler J, Self C. Intra-and interseasonal autoregressive prediction of dengue outbreaks using local weather and regional climate for a tropical environment in Colombia. Am J Trop Med Hyg 2014;91(3):598–610. https://doi.org/10.4269/ajtmh. 13-0303 PMID: 24957546.

[37] Aguiar M, Ballesteros S, Kooi BW, Stollenwerk N. The role of seasonality and import in a minimalistic multi-strain dengue model capturing differences between primary and secondary infections: complex dynamics and its implications for data analysis. J Theor Biol 2011;289:181–96. https://doi.org/10.1016/j.jtbi.2011.08.043 PMID: 21907213.

[38] Johansson MA, Cummings DAT, Glass GE. Multiyear climate variability and dengue—El Nino southern oscillation, weather, and dengue incidence in Puerto Rico, Mexico, and Thailand: a longitudinal data analysis. PLoS Med 2009;6(11):e1000168. https://doi.org/10.1371/journal.pmed.1000168 PMID: 19918363.

[39] Chan EH, Sahai V, Conrad C, Brownstein JS. Using web search query data to monitor dengue epidemics: a new model for neglected tropical disease surveillance. PLoS Negl Trop Dis 2011;5(5):e1206. https://doi.org/10.1371/journal.pntd.0001206 PMID: 21647308.

[40] Althouse BM, Ng YY, Cummings DAT. Prediction of dengue incidence using search query surveillance. PLoS Negl Trop Dis 2011;5(8):e1258. https://doi.org/10.1371/journal.pntd.0001258 PMID: 21829744.

[41] Advances in using internet searches to track dengue Yang S, Kou SC, Lu F, Brownstein JS, Brooke N, Santillana M. PLoS Comput Biol. .

[42] Gubler DJ, Meltzer M. Impact of dengue/dengue hemorrhagic fever on the developing world. Adv Virus Res 1999. Elsevier.

[43] DengueNet. http://www.who.int/csr/disease/Dengue/Denguenet/en/index.html(accessed 7 November 2006).

[44] Focks DA, Alexander N. Multicountry study of Aedes aegypti pupal productivity survey methodology: findings and recommendations (TDR/IRM/DEN/06.1). Geneva: Special Programme for Research and Training in Tropical Diseases (TDR); 2006. 2006.

[45] TDR. Dengue diagnostics: Proceedings of an international workshop. 2005. TDR/IRM/DIAG/DEN/05.1.

[46] Wellcome Trust – TDR. Topics in international health: dengue. CD Rom. Wallingford, Oxford, UK: CABI Publishing; 2005.

[47] Geneva, Switzerland WHO WHO. Scientific working group report on dengue. 2007.

[48] TDR/WHO. Dengue: guidelines for diagnosis, treatment, prevention and control. Geneva, Switzerland: TDR/WHO; 2009.

[49] Guzman MG, Kouri G. Dengue: an update. Lancet Infect Dis 2002;2:33–42.

[50] Gubler DJ. The changing epidemiology of yellow fever and Dengue, 1900 to 2003: full circle? Comp Immunol Microbiol Infect Dis 2004;27:319–30.

[51] WHO scientific working group on dengue; meeting report 3–5 April, 2000. Geneva, Switzerland: WHO; 2000.

[52] WHO. World health assembly resolution WHA55.17, adopted by the 55th world health assembly. Geneva, Switzerland: WHO; 2002 [Dengue fever and Dengue haemorrhagic fever prevention and control].

[53] WHO. Revision of the international health regulations. World health Assembly Resolution WHA58.3, adopted by the 58th world health Assembly. Geneva, Switzerland: WHO; 2005.

[54] Rodriguez MM, Bisset JA, Fernandez D. Levels of insecticide resistance and resistance mechanisms in *Aedes aegypti* from some Latin American countries. J Am Mosq Control Assoc 2007;23:420–9.

[55] Brengues C, et al. Pyrethroid and DDT cross-resistance in *Aedes aegypti* is correlated with novel mutations in the voltage-gated sodium channel gene. Med Vet Entomol 2003;17:87–94.

[56] Murray NE, Quam MB, Wilder-Smith A. Epidemiology of dengue: past, present and future prospects. Clin Epidemiol 2013. https://doi.org/10.2147/CLEP.S34440.

[57] Whitehead SS, et al. Prospects for dengue virus vaccine. Nat Rev Microbiol 2007;518. 528.

[58] Nam VS, Yen NT, Holynska M, Reid JW, Kay BH. National progress in dengue vector control in Vietnam: survey for Mesocyclops (Copepoda), Micronecta (Corixidae), and fish as biological control agents. Am J Trop Med Hyg 2000;62:5–10.

[59] Kay BH, et al. Control of Aedes vectors of dengue in three provinces of Vietnam by use of Mesocyclops (Copepoda) and community-based methods validated by entomologic, clinical, and serological surveillance. Am J Trop Med Hyg 2002;66:40–8.

[60] Heintze C, Garrido MV, Kroeger A. What do community-based dengue control programmes achieve? A systematic review of published evaluations. Trans R Soc Trop Med Hyg 2007;101:317–25.

[61] Kroeger A, et al. Effective control of dengue vectors with curtains and water container covers treated with insecticide in Mexico and Venezuela: cluster randomised trials. Br Med J 2006;332:1247–52.

[62] Mulla MS, Thavara U, Tawatsin A, Kong-Ngamsuk W, Chompoosri J. Mosquito burden and impact on the poor: measures and costs for personal protection in some communities. J Am Mosq Control Assoc 2001;17:153–9.

[63] Hombach J. Vaccines against dengue: a review of current candidate vaccines at advanced development stages. Rev Panam Salud Pública 2007;21:254–60.

[64] Hombach J, Cardosa JM, Sabchareon A, Vaughn DW, Barrett ADT. Scientific consultation on immunological correlates of protection induced by Dengue vaccines. Report from a meeting held at the World Health Organization 17–18 November 2005. Vaccine 2007;25:4130–9.

[65] Bettramello M, et al. The human immune response to dengue virus is dominated by highly cross-reactive antibodies endowed with neutralizing and enhancing activity. Cell Host Microbe 2010;8:271–83.

[66] Dejnirattisai W, et al. Cross-reacting antibodies enhance Dengue virus infection in humans. Science 2010;328:745–8.

[67] Rothman AL. Dengue: defining protective versus pathologic immunity. J Clin Invest 2004;113:946–51.

[68] Guirakhoo F, et al. Live attenuated chimeric yellow fever dengue type 2 (ChimeriVax-DEN2) vaccine: phase I clinical trial for safety and immunogenicity: effect of yellow fever pre-immunity in induction of cross neutralizing antibody responses to all 4 Dengue serotypes. Hum Vaccine 2006;2:60–7.

[69] Durbin AP, et al. rDEN4 Delta 30, a live attenuated Dengue virus type 4 vaccine candidate, is safe, immunogenic, and highly infectious in healthy adult volunteers. J Infect Dis 2005;191:710–8.

[70] Raviprakash K, et al. A chimeric tetravalent dengue DNA vaccine elicits neutralizing antibody to all four virus serotypes in rhesus macaques. Virology 2006;353:166–73.

[71] Hermida L, et al. A recombinant fusion protein containing the domain III of the dengue-2 envelope protein is immunogenic and protective in nonhuman primates. Vaccine 2006;24:3165–71.

[72] Whitehead SS, et al. A live, attenuated Dengue virus type 1 vaccine candidate with a 30-nucleotide deletion in the 3′ untranslated region is highly attenuated and immunogenic in monkeys. J Virol 2003;77:1653–7.

[73] Edelman R, et al. Phase I trial of 16 formulations of a tetravalent live-attenuated dengue vaccine. Am J Trop Med Hyg 2003;69:48–60.

[74] Wright PF, et al. Phase 1 trial of the dengue virus type 4 vaccine candidate rDEN4Δ30-4995 in healthy adult volunteers. Am J Trop Med Hyg 2009;81:834–41.

[75] Morrison D, et al. A novel tetravalent dengue vaccine is well tolerated and immunogenic against all 4 serotypes in flavivirus-naive adults. J Infect Dis 2010;201:370–7.

[76] Parks W, Lloyd L. Planning social mobilization and communication for Dengue fever prevention and control: a step-by-step guide. Geneva, Switzerland: WHO; 2004.

[77] Andre Sofair. FDA approved vaccine for prevention of Dengue disease in endemic regions. FDA news room; May 2019.

Chapter 5

# Mosquito-borne diseases

Muhammad A. Saleem[1], Iryna Lobanova[2,3]
[1]*Resident Physician, Family Medicine, Mercyhealth, Janesville, WI, United States;* [2]*Project Manager, Dengue virus disease project, Zeenat Qureshi Stroke Institute, St. Cloud, MN, United States;* [3]*University of Missouri, Columbia, MO, United States*

## The dangerous insects

Mosquitoes may lack in size but are by far the deadliest living being on earth beating grizzly bears, tigers, cobras, and hippos with huge margins. These pesky insects are capable of killing more people in one day than sharks do in a century. Unarguably, they are able to carry and spread more potentially fatal diseases than any other vector on the planet. The human lives these blood-thirsty creatures take and the resultant lost productivity accounts to billions of dollars annually. Mosquitoes not only pose a deadlier threat to humans but also to larger number of animals including amphibians, reptiles, squirrels, rabbits, and small mammals. The cross-infections caused by them have resulted in novel epidemics in recent decades. Investigations have revealed their remarkable ability to adapt to new environments, quick and large scale reproduction, resistance to certain insecticides, and alteration in feeding habits to survive against control methods.

## Life-cycle of mosquitoes

The mosquitoes lay plentiful eggs and the bloodsucking insects can grow from an egg to an adult in just five days. After the female mosquito obtains blood meal, she lays several hundred eggs directly on or near water, soil, and at the base of some plants in places that may fill with water. The eggs can survive in desiccated conditions for a few months. The size of their nursery can be as small as a jar top. The eggs hatch when exposed to water and the length of time it takes for them to hatch also depends on water temperature and the type of mosquito. The egg hatches into bristly aquatic larva about 8 mm long. These larvae swim with a jerking, wriggling movement thereby called "wrigglers." Larva mostly feed on algae and organic debris from sewage. Larva molts several times and most species surface to breathe air before developing into

Dengue Virus Disease. https://doi.org/10.1016/B978-0-12-818270-3.00005-9

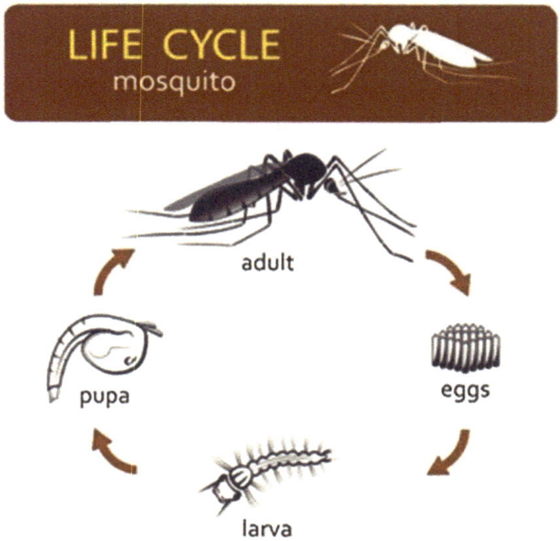


FIGURE 5.1 Lifecycle of mosquito.

pupa. Mosquitoes in their pupal stage are called tumblers and are actively free-swimming thereby called "tumblers." The pupa also lives in the water but no longer feeds. After almost a week the pupa transforms into the adult mosquito (see Fig. 5.1). The duration of lifecycle may range from 4 days to a month depending on the species of mosquito. The adult mosquito emerges onto the water's surface and flies away, ready to begin its lifecycle [1].

Mosquitoes can fly long distances; some more than 20 miles from the water source that produced them. But they do not fly fast, only about 4 miles an hour. Since they typically fly into the wind to help detect host odors, fewer mosquitoes are about on windy days.

## The sniffy navigation

The mosquitoes use scent to find humans. The chemistry of our perspiration especially the lactate is attractive to them. They can also sense carbon dioxide in our exhalation sand follow the stream back to our faces. Mosquitoes can also detect body heat and movements to identify the potential target [2]. The frequency of wing beats gives them their unique hum and may serve as a means of gender recognition [3].

## Common genus of mosquitoes

There are thousands of different species of mosquitoes. While the deadliest types feed on our blood and spread horrible diseases, some of them are parasites on other animals and rare mosquito species do not feed on blood at all. Here we discuss the most common genus of mosquitoes.

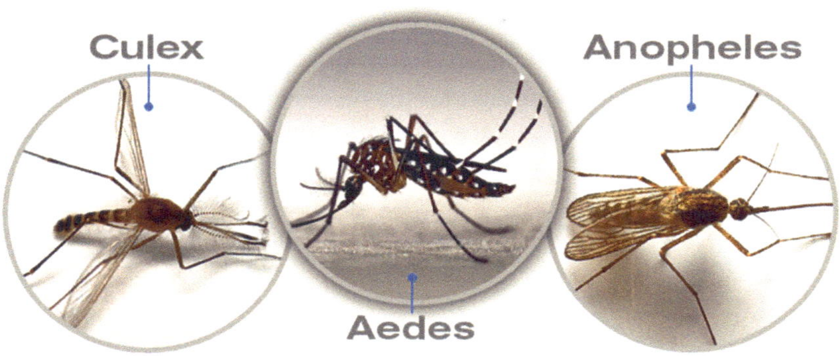


## Aedes

This group includes many species that transmit disease to humans. These include inland flood water mosquito (*Aedes vexans*), the Asian tiger mosquito (*Aedes albopictus*) and the tree hole mosquito (*Ochlerotatus triseriatus*)—all of which feed on the blood of mammals. Flood water mosquito lay its eggs on soil which becomes flooded thereby allowing the life cycle to proceed. Asian tiger and Tree hole mosquitoes breed in containers, laying their eggs in small water filled cavities and thrive in urban areas.

Inland flood water mosquitoes are brown with pale B-shaped marks on their abdomens. Typical outbreaks are seen when river backwaters and low-lying places become flooded. They are usually the first crop noticed in spring and later after heavy rainfall. They can fly more than 10 miles from their breeding places in search of blood meals. They are usually active in the dark and typically the infestations die in autumn with the first frosting.

The Asian tiger mosquito is black with one white "racing" stripe on its thorax. It is usually active during day time and is capable of carrying Zika, Dengue, Chikungunya, West Nile viruses and Yellow fever viruses. The biggest vector of La Crosse encephalitis is the Tree hole mosquito. It is a dark brown mosquito with silvery white bands on legs. It bites by day and lays its eggs in small containers where water will pool, such as tree holes, discarded tires, cans, buckets, and barrels. They often are found in and around wooded areas.

*Aedes aegypti*, the Yellow fever mosquito, is closely associated with and feeds primarily on humans. It is a small container breeder and will be collected around and in homes. It is active primarily during dusk and dawn. It is often noticed flying about ankles looking for an opportunity to feed. It also wins the distinction of main vector for Dengue virus, Chikungunya virus, and Zika virus.

## Anopheles

Also known as marsh mosquitoes, this genus has 460 different species. The most prominent species in *Anopheles gambiae* which is well known for carrying malaria in Africa and *Anopheles freeborni* in North America. They breed in natural water collections and therefore breeding increases in the rainy season when water collects in bottles, tins, buckets, tyres etc. Construction sites provide ample breeding places for the mosquito. Anopheles mosquitoes enter the house near sunset and again in early hours of morning. They start biting by late evening and the peak of biting activity is at midnight and early hours of morning. Mosquitoes can fly up to several kilometers and they can reach far off places by taking shelter in motor vehicles, ships, and aircraft.

## Culex

Also known as the house mosquito, the Culex genus contains several species that are vector of West Nile virus, West Nile encephalitis, and Rift Valley fever. These include *Culex pipiens*, *Culex quinquefasciatus*, and *Culex tarsalis*. They typically bite at dusk and after dark. By day, they rest in and around structures and vegetation. They lay eggs on still water in natural and manmade containers. Adult *Culex* mosquitoes do not fly far from where they develop as larvae. Unlike other mosquitoes that die with the coming of the first hard frost in autumn, the house mosquito can "over-winter" in protected places like sewers, crawlspaces, and basements.

## Culiseta

These are larger mosquitoes which bear superficial resemblance to Culex. They have adapted to the cold, and are found everywhere except South America. The larvae of most species are found in ground waters such as bogs, marshes, ponds, streams, ditches, and rock pools, but an African species occurs in tree holes, a common eastern Palearctic species occurs in wells and rock pools, and several Australian species occur underground.

## Mansonia

These mosquitoes are bigger than most, and are black or brown with sparkling wings and legs. They are found in most parts of the world, and are known to transmit encephalitis. *Mansonia* mosquitoes prefer to bite in the evening. They breed in ponds and lakes containing certain aquatic plants, especially the floating type like *Pistia stratiotes* and water hyacinth. The eggs are laid in star-shaped clusters on the undersurface of leaves of these plants. The larvae and pupae are found attached to the rootlets of these plants by their siphon tubes. They obtain their air supply from these rootlets. When about to become adult, these pupae come to the surface of water and the fully formed adults emerge and escape.

### Psorophora

With a realm of species, *Psorophora* mosquitoes vary in size, from small to very large. They are mostly found in tropical areas of North and South America, and are vectors for Ilheus virus and encephalitis. They are sometimes referred to as flood water mosquitoes, since they like laying their eggs on mud.

### Toxorhynchites

Also known as the elephant mosquito, this genus does not consume blood. Like their male counterparts, the females feed on plant nectar, and do not pose a risk to humans. Interestingly, their larvae prey on the larvae of other mosquitoes; it is been suggested that *Toxorhynchites* could be introduced in areas they do not generally inhabit to help fight Dengue viral infection.

### Wyeomyia

Mostly found in Central and South America, this genus are not known to carry diseases, so do not pose a risk to humans. There are 140 known species, and they generally inhabit flowers, bamboo, tree holes, and containers. Adults are active during the day, usually near larval habitats. Some species are found at characteristic elevations in the forest canopy, with others appearing to be restricted to ground level.

## Techniques for eradication

Integrated pest management is a science-based, common-sense approach for managing mosquitoes. It uses a variety of pest management techniques that focus on pest prevention, pest reduction, and the elimination of conditions that lead to pest infestations [4]. Such pest management programs also rely heavily on resident education and pest monitoring. Integrated pest management uses various ways to control mosquito populations with decision based on surveillance to keep track of infestations. This is backed with appropriate use of insecticides. The cornerstones of such a management strategy are elaborated below:

### Remove mosquito habitats

Since most species of mosquitoes lay their eggs in or near water, it is an important part of mosquito control strategy to make sure that there are no potential reservoirs. The mosquitoes love standing water and would rush to make it their home. The obvious ones are puddles and small ponds but small water bodies that are usually overlooked include landscape ponds, lawn or yard ornaments, puddles, bird baths, clogged gutters, and downspouts [5]. It is important to drain rain gutters, old tires, plastic covers, and toys where mosquitoes may be hiding. Draining temporary pools of water or filling them with dirt and keeping swimming pool water treated are also cornerstone approaches to handling potential mosquito breeding sites in a modern household.

## Control mosquitoes at the larval stage

Although the most natural solution to eliminate mosquito reproduction is to drain their breeding grounds, sometimes it is very difficult. Mosquitoes have been demonstrated to maintain their colonies in as much as half an inch bodies of water. Areas like runoff drains and landscaping are inherently designed to hold water. The greatest impact on mosquito occurs when they are immobile, accessible, concentrated, and fragile. Larviciding typically involves applying pesticides containing methoprene or *Bacillus thuringiensis israelensis* or *B. sphaericus* bacteria, to water where mosquito larvae develop. As larvae feed on Bacillus, a bacterial toxin is released which perforates mosquito's gut. Methoprene is a larvae growth regulator and works by disrupting larvae metamorphosis [6]. The toxicity of both of these methods is very low and they have been deemed safe to be used in waters containing fish. They can be used effectively where it is undesirable or impractical to empty the water in containers, such as water in decorative pools or horse watering tanks. Other types of larvicides include those that cover the surface of the water with thin films of liquid designed to prevent larvae from obtaining oxygen at the water's surface.

## Control adult mosquitoes

Adult mosquito control involves application of fine droplets of pesticides released as an ultralow volume treatment from a specialized truck or aerial equipment. This technique is usually used to combat an outbreak of mosquito-borne diseases being transmitted by adult mosquitoes. In contrast to larviciding, adulticiding is a broad spectrum application that can kill beneficial insects as well and is much more expensive. Adulticiding should be seen as a supplement to larviciding, to be used when mosquitoes become too numerous or when high levels of virus activity in mosquitoes threaten populated areas with disease [7].

## Use of structural barriers

Since the mosquitoes frequently bite indoors, using structural barriers is an important way to reduce the incidence of bites. It is important to make sure that the windows and door are bug tight by maintaining screens integrity and sealing all gaps through which mosquitoes might enter. Completely covering baby carriers and beds with netting. Nets can be especially important for protecting a sick person from getting more mosquito bites, which could transmit the disease to other people. Repellents like N,N-diethyl-meta-toluamide, picaridin and lemon oil of eucalyptus provide fair benefit when used appropriately [8].

## Mosquitoes in America

In the 17th century, the trade ships arriving at the Southern ports in the United States are believed to have brought *Aedes aegypti* mosquito and resultantly Yellow fever and Dengue fever to the country [9,10]. Initially, the epidemics were confined to the temperate and tropical zones in south and the west [11], until the 19th century numerous outbreaks of malaria have been recorded as far north as Massachusetts [12]. Malaria spread quickly in the Midwest states along the Mississippi valley during the American Revolutionary War and the Civil War. Malaria emerged as a major catastrophe during 4 years of the Civil War claiming thousands of lives [13]. The transmission of disease causing organism was proven in 1889 and since then the US government has undertaken multidisciplinary efforts to control mosquitoes [14]. With changes in the lifestyle and population spending more time indoors than outdoors, frequent use of door and window screens, depopulation of the rural south, and improved socioeconomic conditions malaria was no longer endemic in any area of continental United States by the early 1940s [15].

During the 18th and beginning of 19th century, Yellow fever epidemics were common in Northeastern states [16]. It is believed that the Yellow fever epidemic in Philadelphia in 1793 was a contributing factor in the decision to move the United States capital to Washington [17]. Between 1693 and 1905, approximately 10 million deaths are attributed to Yellow fever along the Mississippi River, from the Gulf of Mexico to Memphis, Tennessee, St. Louis, Missouri, and New Orleans [18]. The epidemics that claimed most of the lives occurred during summer in port cities with active trade with the Caribbean Islands. After identification of *Aedes aegypti* as the primary vector of yellow fever in 1901, the US companies that took over the construction of Panama Canal participated actively in the mosquito eradication drives by draining or oiling the larva breeding sites and fumigation [19]. Later on, the programs implemented by Pan American Health Organization eradicated the *Aedes aegypti* mosquito and consequently Dengue viral infection until 1970. Because of the questions raised about the safety of Dichlorodiphenyltrichloroethane use in environment and the huge success of elimination program, the elimination programs fell on the back foot leading to reemergence of mosquito, escalation in its population and ultimately failure to eradicate Dengue viral infection [20].

The socioeconomic changes in the United States spanning on 18th and 19th century have substantially affected the transmission of those pathogens for which humans are the primary amplifying host. The availability of potable water has eliminated the need to store water in the household and thereby reduced the breeding grounds of mosquitoes. Likewise, improved sanitation and sewage disposal processing have shrunk the vector nurseries as well. After the second world war, the boom in the US economy improved the living standards and aided widespread use of television and air-conditioning [21]. These commodities encouraged general population to spend more time in indoor screened areas thereby reducing potential mosquito targets [22].

The improvement in socioeconomic conditions has lead to substantial reduction in prevalence and transmission of anthroponotic pathogens like dengue and malaria viruses but it does not affect the transmission of zoonotic pathogens such as West Nile virus, St. Louis encephalitis, Eastern equine encephalitis, Western equine encephalitis, and La Crosse encephalitis [23]. Since animals other than humans play a major role in their transmission, these are maintained in natural transmission cycles without being affected by improved accommodation for humans. Therefore, zoonotic viruses like eastern equine encephalitis and La Crosse viruses continue to cause disease throughout the continental United States, Southern Canada, and South America [24].

## West Nile virus

This virus first appeared in the United States in 1999 and spread by 2001 to Florida. This disease is primarily an infection found in birds, but can also be transferred to humans, dogs, horses, and other animals. Humans, while also susceptible, often show no symptoms (less than 1% of those infected show symptoms). The mortality rate of those who become ill ranges from 3% to 15%, and is highest among the elderly.

West Nile virus is an arbovirus of the *Flavivirus* kind in the family Flaviviridae. Mosquitoes spread from West Nile to 48 of the 50 US states, Africa, Europe, the Middle East, and West and Central Asia. In 2012, the US experienced one of its worst epidemics where 286 people died, with the state of Texas being hit hardest [25]. As of 2014, there have been 36,437 cases of West Nile virus. Of these, 15,774 have resulted in meningitis/encephalitis and 1538 were fatal. There have been at least 1.5 million infections (82% are asymptomatic) and over 350,000 cases of West Nile fever, but the disease is underreported due to its similarity to other viral infections. This virus usually circulates between mosquitoes and birds in Africa and Europe. However, in 1999 an outbreak of West Nile encephalitis was reported in New York City. Since then, the virus has spread to 48 states and the District of Columbia [26–30].

Most people who get West Nile virus do not have any symptoms. About 1 in 5 will have a fever and other flu-like symptoms. A few people (less than 1%) get more serious infection West Nile neuroinvasive disease, which causes meningitis, encephalitis, meningoencephalitis, and poliomyelitis-like syndrome. People of advanced age, the very young, or those with immunosuppression are most susceptible. Many patients with WNND have normal neuroimaging studies, although abnormalities may be present in various cerebral areas including the basal ganglia, thalamus, cerebellum, and brainstem.

West Nile virus encephalitis is the most common neuroinvasive manifestation of West Nile neuroinvasive disease. West Nile virus encephalitis presents with similar symptoms to others viral encephalitis with fever, headaches, and altered mental states. A prominent finding in West Nile virus encephalitis is the muscular weakness (30%–50% of patients with encephalitis), often with lower motor neuron symptoms, flaccid paralysis, and hyporeflexia with no sensory abnormalities. West Nile meningitis usually involves fever, headache, and stiff neck. Pleocytosis, an increase of white blood cells in cerebrospinal fluid, is also present. Changes in consciousness are not usually seen and are mild when present. West Nile poliomyelitis, an acute flaccid paralysis syndrome associated with infection, is less common than West Nile meningitis or West Nile virus encephalitis. This syndrome is generally characterized by the acute onset of asymmetric limb weakness or paralysis in the absence of sensory loss. The pain sometimes precedes the paralysis. The paralysis can occur in the absence of fever, headache, or other common symptoms associated with West Nile virus infection. Involvement of respiratory muscles, leading to acute respiratory failure, can sometimes occur. West Nile reversible paralysis, like West Nile poliomyelitis, the weakness or paralysis is asymmetric [31–36]. Reported cases have been noted to have an initial preservation of deep tendon reflexes, which is not expected for a pure anterior horn involvement. The disconnect of the upper motor neuron influences on the anterior horn cells possibly by myelitis or glutamate excitotoxicity have been suggested. The prognosis for recovery is excellent [37].

Nonneurologic complications of West Nile virus include fulminant hepatitis, pancreatitis [27], myocarditis, rhabdomyolysis [34], orchitis [38], optic neuritis [31], cardiac dysrhythmias, and hemorrhagic fever with coagulopathy [38], chorioretinitis [39,40]. The real danger may be for pregnant women and their babies. It is linked to a birth defect called microcephaly [32]. There is no specific treatment for West Nile virus [41,42].

## Chikungunya virus

Chikungunya is another arbovirus transmitted through the bite of an infected female mosquito *Ae. aegypti* and, to a lesser extent, the *Ae. albopictus* species. Chikungunya virus has recently appeared in such places as India, Sri Lanka, Mauritius, and countries in Europe involved in frequent tourism to these destinations. Concern has arisen recently that will soon be increasing the range in Europe over the spread of the Asian Tiger mosquitoes (*Aedes albopictus*), that can act as significant vectors for this infection. The traditional range for this virus also includes Africa and South East Asia. Infection with chikungunya can be severe and temporarily debilitating, but is generally not life-threatening in otherwise healthy people. No vaccine or curative drug treatment is currently available. Prevention must rely entirely on measures that reduce exposure to mosquito bites [43–49].

Typical symptoms of infection by Chikungunya virus are:

- Abrupt onset of fever above 38°C lasting 2–3 days
- Moderate headache
- Moderate-to-intense joint and tendon pain and swelling
- Intense muscle pain
- Fatigue
- Rash (between the 2nd and 5th day in 50% of the cases)
- Conjunctivitis/eye redness (30% of the cases)
- Very rare cases include neurological manifestations in the form of encephalitis, Guillain-Barre syndrome, and myelitis among other complications.

No specific treatment for Chikungunya viral fever is available at this stage. The infected person should take analgesics to treat fever (except acetylsalicylic acid) and antiinflammatory for the joint pain, drink plenty of fluids, rest and eat normally, and take precautions against mosquito biting to prevent spread of disease [43–49].

Symptoms usually appear 3–7 days after the bit of an infected mosquito. The most common symptoms of chikungunya virus infection are fever and joint pain. Other symptoms may include a headache, muscle pain, joint swelling, or rash. Chikungunya viral disease does not often result in death, but the symptoms can be severe and disabling. Most patients recover fully, but in some cases, joint pain may persist for several months or even years. Occasional cases of eye, neurological, and heart complications have been reported, as well as gastrointestinal complaints. Serious complications are not common except in older people where the disease can cause of death. Often symptoms in infected individuals are weak and the infection may go unrecognized or be misdiagnosed in areas where Dengue viral illness occurs [43–49]. There is no vaccine, and primary treatment is limited to pain medication [50].

## Eastern equine encephalitis virus

Eastern equine encephalitis virus is transmitted to humans by the bite of an infected mosquito. Most cases occur in the Atlantic and Gulf Coast states. Most persons infected with Eastern equine encephalitis virus have no apparent illness. Severe cases of Eastern equine encephalitis begin with the sudden onset of a headache, high fever, chills, and vomiting. The illness may then progress into disorientation, seizures, or coma. Eastern equine encephalitis is one of the most severe mosquito-transmitted diseases in the United States with approximately 33% mortality and significant brain damage in most survivors. There is no specific treatment for Eastern equine encephalitis; care is based on symptoms [51,52].

This is a disease found in horses, but there is a vaccine available. The virus is maintained in birds, which are not affected by the disease. Humans do sometimes contract the disease, but humans are not a preferred target of the mosquitoes carrying this virus and are considered a "dead end" host. This means that, while a human can contract Eastern equine encephalitis, mosquitoes have not been shown to become infected by a human host and spread it to other organisms. There are normally a few cases of Eastern equine encephalitis in Florida each year and the disease has shown to have a relatively high mortality rate of approximately 35%. Many of those who survive Eastern equine encephalitis will have mild to severe brain damage resulting from high fever.

## Japanese encephalitis virus

Countries with the proven epidemics of Japanese encephalitis virus are India, Pakistan, Nepal, Sri Lanka, Burma, Laos, Vietnam, Malaysia, Singapore, Philippines, Indonesia, China, maritime Siberia, Korea, and Japan. The virus is transmitted to humans by the bite of an infected mosquito, which serves as a dead end host due to its short duration and low viremia in man. Most important mosquito vector in Asia is *Culex tritaeniorhynchus,* which breeds in the stagnant water like paddy fields or drainage ditches. Other species are *Culex vishnui* (India), *C. gelides, C. fusco cephalea* (India, Malaysia, Thailand), and *C. pipiens.* About 50%–60% of the survivors suffer from serious long-term neurologic complications manifesting as convulsions, tremors, paralysis, ataxia, memory loss, impaired cognition, behavioral disturbance, and other such symptoms. There is an incubation period of 4–14 days in humans during Japanese encephalitis virus infection and patients are presented with few days of fever including coryza, diarrhea, and rigors. Convulsions occur 10% more frequently in children (85% of cases) than in adult patients (75% of cases). Vector control alone cannot be relied upon to prevent Japanese encephalitis virus since it is practically almost impossible to control mosquito density in the rural areas which are the worst affected areas due to poor socioeconomic conditions. Three types of Japanese encephalitis virus vaccine are currently in use: mouse-brain derived inactivated, cell-culture-derived inactivated, and cell-culture-derived live attenuated Japanese encephalitis virus vaccine. Formalin-inactivated vaccines are safe and effective against Japanese encephalitis virus for at least 30 years [53–55].

Japanese encephalitis virus is the leading cause of vaccine-preventable encephalitis in Asia and the Western Pacific [53–55]. Pigs are an important maintenance host for this virus, which is mainly transmitted by night-biting mosquitoes in the *Culex tritaeniorhynchus* group. Japanese encephalitis virus is found in India and South East Asia upto Japan and has a similar risk for fatality and permanent debilitation as Eastern Equine Encephalitis, except Japanese encephalitis virus, affects a wider range of age groups. It has recently expanded its range to reach northern Australia. A total of 30% of those who show Japanese encephalitis virus symptoms die and another 30% develop serious and permanent neurological damage. Between 30,000–50,000 clinical infections are reported each year across Asia. The majority of infections, however, are asymptomatic [56–60]. Steps to prevent Japanese encephalitis virus include using personal protective measures to prevent mosquito bites. As previously mentioned, an effective vaccine is available for preventing this infection [61].

## Murray valley encephalitis virus

Murray Valley encephalitis is an uncommon disease caused by the MVE encephalitis virus. Murray Valley encephalitis virus is a flavivirus endemic to northern Australia and Papua New Guinea [62]. The most important species of mosquito to carry the virus is the common banded mosquito, *Culex annulirostris*. Most people with this infection remain completely well while others may only develop a mild illness with fever. A small proportion of those infected develop encephalitis.

Symptoms of Murray Valley encephalitis usually appear 5–28 days (average 14 days) after the infected mosquito bite. The early symptoms include the following: a headache, fever, nausea, vomiting, and muscle aches. Symptoms may also include drowsiness, confusion, seizures or fits (especially in infants), and in severe cases delirium, coma, and death. Some who recover are left with ongoing problems such as deafness or epilepsy. There is no specific treatment for Murray Valley encephalitis [63–65].

## La Crosse encephalitis virus

La Crosse encephalitis was discovered in 1965, after the virus was isolated from preserved brain tissue and spinal cord of a child who died from the unknown infection in La Crosse, Wisconsin in 1960. It occurs in the Appalachian and Midwestern regions of the United States. Recently there has been an increase of cases in the South East of the United States. An explanation to this may be that the mosquito *Aedes albopictus* is also an efficient vector of La Crosse virus [66]. Between 2004 and 2013 there were 787 total cases of La Crosse encephalitis and 11 deaths in the U.S [67].

Many people infected with La Crosse encephalitis virus have no visible symptoms. Among people who become ill, initial symptoms include fever, headache, nausea, vomiting, and tiredness. Some of those who become ill develop the severe neuroinvasive disease. The incubation period for La Crosse encephalitis virus disease ranges from 5 to 15 days. La Crosse encephalitis virus disease is usually characterized by fever (usually lasting 2–3 days), headache, nausea, vomiting, fatigue, and lethargy. The severe La Crosse encephalitis virus disease often involves encephalitis and can include seizures, coma, and paralysis. Severe disease occurs most often in children under the age of 16 years. In rare cases, long-term disability or death can result from La Crosse encephalitis. No vaccine against La Crosse encephalitis virus infection or specific antiviral treatment for clinical La Crosse encephalitis virus infection is available. Patients with suspected La Crosse encephalitis virus encephalitis should be hospitalized, appropriate serologic and other diagnostic tests performed, and supportive treatment (including seizure control) provided [68–70].

## Malaria

Malaria is a disease transmitted by *Anopheles* mosquitoes in the genus distributed throughout the world as a mosquito-borne disease caused by a parasite. Malaria is present in more than 100 countries and imposes an economically significant burden on the populations of at least 80 million. Malaria kills at least 1.1 million people per year, and probably more due to incomplete reporting in many of the countries on which it imposes the greatest burdens. Four species of parasites affect humans, but two of them, *Plasmodium falciparum* and *P. vivax* account for more than 95% of malaria cases. *P. falciparum,* the maximum dangerous of the pair which exists throughout the deep tropics from Africa to Asia and South America. In 2015, an estimated 214 million cases of malaria occurred worldwide and 438,000 people died, mostly children in the African Region. Malaria was endemic in the United States in the 19th and 20th centuries. About 1500 cases of malaria are diagnosed in the United States each year. The vast majority of cases in the United States are in travelers and immigrants returning from countries where malaria transmission occurs, many from sub-Saharan Africa and South Asia [71–76].

The immune evasiveness of malaria parasites prevents complete immunity from developing, but older children and adults who have experienced multiple infections, enjoy some level of protection from the most severe manifestations of the illness.

Certain complications, such as cerebral malaria, strike quickly, clogging small blood vessels in the brain to produce coma. Stories of expatriates falling ill on a Friday, putting off treatment till Monday, and dying over the weekend are not uncommon. Thus, malaria prevention requires serious attention when visiting areas where it is transmitted. Although no vaccine is currently available, prophylactic drugs and measures that reduce exposure to night-biting *Anopheles* mosquitoes, such as bed nets and repellents can be very effective. Unlike some infections, the victims of malaria often never get a second chance.

The causative agent of tropical malaria-infected red blood cells, especially mature trophozoites, adhere to the vascular endothelium of venular blood vessel walls and do not freely circulate in the blood. When this sequestration of infected erythrocytes occurs in the vessels of the brain, it is believed to be a factor in causing the severe disease syndrome known as cerebral malaria, which is associated with high mortality [77–80].

## Uncomplicated malaria

The classical (but rarely observed) malaria attack lasts 6–10 h. It consists of a cold stage (sensation of cold, shivering), a hot stage (fever, headaches, vomiting; seizures in young children), and finally a sweating stage (sweats, return to normal temperature, tiredness). Classically (but infrequently observed) the attacks occur every second day with the "tertian" parasites (*P. falciparum*, *P. vivax*, and *P. ovale*) and every third day with the "quartan" parasite (*P. malaria*). More commonly, the patient presents with a combination of the following symptoms: fever, chills, sweats, headaches, nausea, vomiting, body aches, and general malaise.

The diagnosis of malaria depends on the demonstration of parasites in the blood, usually by microscopy. Additional laboratory findings may include mild anemia, the slight decrease in blood platelets (thrombocytopenia), elevation of bilirubin, and elevation of aminotransferases [81–89].

## Severe malaria

Severe malaria occurs when infections are complicated by serious organ failures or abnormalities in the patient's blood or metabolism. The manifestations of severe malaria include the following:

- Cerebral malaria, with abnormal behavior, impairment of consciousness, seizures, coma, or other neurologic abnormalities
- Severe anemia due to hemolysis
- Hemoglobinuria due to hemolysis
- Acute respiratory distress syndrome, an inflammatory reaction in the lungs that inhibits oxygen exchange, which may occur even after the parasite counts have decreased in response to treatment
- Abnormalities in blood coagulation
- Low blood pressure caused by cardiovascular collapse
- Acute kidney failure
- Hyperparasitemia, where more than 5% of the red blood cells are infected by malaria parasites
- Metabolic acidosis, often in association with hypoglycemia
- Hypoglycemia, which may also occur in pregnant women with uncomplicated malaria, or after treatment with quinine
- Neurologic defects may occasionally persist following cerebral malaria, especially in children. Such defects include ataxia, palsies, speech difficulties, deafness, and blindness [90–95].

Quinine and other antimalarial drugs cure patients by attacking the parasites in the blood. Despite the need, no effective vaccine exists. The recommended treatment for malaria is the combination of antimalarial medications that includes an artemisinin. The second medication may be either mefloquine, lumefantrine, or sulfadoxine pyrimethamine. Quinine along with doxycycline may be used if artemisinin is not available. There are a number of medicines that can help prevent or interrupt malaria in travelers to areas where the infection is common. Many of these drugs are also used in treatment. Chloroquine may be used where chloroquine-resistant parasites are not common. In places where *Plasmodium* is resistant to one or more medications, three medications:Mefloquine (*Lariam*), doxycycline (available generically), or the combination of atovaquone and proguanil hydrochloride (*Malarone*)—are frequently used when prophylaxis is needed. Doxycycline and the atovaquone plus proguanil combination are the best tolerated; mefloquine is associated with death, suicide, and neurological and psychiatric symptoms [96–100]. Drug resistance is now common against all classes of antimalarial drugs apart from artemisinins [101].

## St. Louis encephalitis virus

This viral disease is maintained in birds and transmitted to both humans and horses by the mosquito *Culex nigrapalpus*. While the virus is present in the wild every year, it occurs as a public health outbreak only periodically. This irregular outbreak pattern is the result of the number of susceptible individuals in the bird population coupled with large populations of the transmitting mosquitoes. These mosquito population increases are triggered by a combination of climactic conditions including drought and rain cycles and individual rain events.

Unfortunately, an accurate prediction model has yet to be developed. Currently, the best method of determining the presence of active viral transmission is the use of sentinel chicken flocks. This information coupled with mosquito population surveillance provides a basis for treating to limit this disease. Humans exposed to the virus often exhibit no symptoms. Those exhibiting symptoms will often suffer long-term neurological impairment. This disease has a mortality rate of 3%–30% and is highest in the elderly population.

Saint. Louis encephalitis is transmitted from birds to humans and other mammals by infected mosquitoes (mainly some *Culex* species). Saint. Louis encephalitis is found throughout the United States, but most often along the Gulf of Mexico, especially Florida. Symptoms are similar to those seen in Eastern Equine encephalitis and like Eastern Equine encephalitis, and there is no vaccine. Mississippi's first case of SLE since 1994 was confirmed in June 2003. Previously the last outbreak of Saint. Louis encephalitis in Mississippi was in 1975 with over 300 reported cases. It was the first confirmed mosquito-borne virus in the United States in 2003. Another occurance turned up in October 2003 in California Riverside County in sentinel chickens. The last

Saint. Louis encephalitis human case in California occurred in 1997 [102–105].

Less than 1% of Saint. Louis encephalitis virus infections are clinically apparent and the vast majority of infections remain undiagnosed. The incubation period for Saint. Louis encephalitis ranges from 5 to 15 days. The onset of illness is usually abrupt, with fever, headache, dizziness, nausea, and malaise. Signs and symptoms intensify over a period of several days to a week. Some patients spontaneously recover after this period, and others develop signs of central nervous system infections, including stiff neck, confusion, disorientation, dizziness, tremors, and unsteadiness. Coma can develop in severe cases. The disease is generally milder in children than in older adults. About 40% of children and young adults with Saint. Louis encephalitis virus disease develop only fever and headache or aseptic meningitis. Almost 90% of elderly persons with Saint. Louis encephalitis virus disease develop encephalitis. The overall case-fatality ratio is 5%–15%. The risk of fatal disease also increases with age [103,106]. There is no vaccine or any other treatments specifically for Saint Louis encephalitis virus, although one study showed that early use of interferon-alpha 2b may decrease the severity of complications [107].

## Yellow fever virus

Yellow fever virus, which has a 400-year history, is found in tropical and subtropical areas in South America and Africa. Every year about 200,000 cases occur with 30,000 deaths in 33 countries. In 2002, one fatal yellow fever death occurred in the United States in an unvaccinated traveler returning from a fishing trip to the Amazon. Like Dengue viral illness, it is transmitted by *Aedes* mosquitoes, especially *Aedes aegypti*, the Yellow Fever mosquito [108,109].

Most people infected with the virus of Yellow Fever have no illness or only mild illness. In persons who develop symptoms, the incubation period (time from infection until illness) is typically 3–6 days. The initial symptoms include sudden onset of fever, chills, severe headache, back pain, general body aches, nausea, vomiting, fatigue, and weakness. Most people improve after the initial presentation. After a brief remission of hours to a day, roughly 15% of cases progress to develop a more severe form of the disease. The severe form is characterized by high fever, jaundice, bleeding, and eventually shock and failure of multiple organs.

Yellow fever disease is diagnosed based on symptoms, physical findings, laboratory testing, and travel history, including the possibility of exposure to infected mosquitoes [110–113]. There is no specific treatment for Yellow fever. Vaccination is highly recommended as a preventive measure for travelers to and people living in endemic countries [114].

## Rift valley fever virus

The disease was first reported among livestock in Rift Valley of Kenya in the early 1900s, and the virus was first isolated in 1931. Outbreaks usually occur during periods of increased rain, which increase the number of mosquitoes. The virus is transmitted through mosquito vectors, as well as through contact with the tissue of infected animals. Two species—*Culex tritaeniorhynchus* and *Aedes vexans*—are known to transmit the virus. The mild symptoms may include fever, muscle pains, and headaches which often last for up to a week. The severe symptoms may include the following: loss of the ability to see (beginning 3 weeks after the infection), infections of the brain that cause severe headaches and confusion, and bleeding together with liver problems, which may occur within the first few days. Those who have bleeding symptoms have a chance of death as high as 50%. Diagnosis relies on viral isolation from tissues or serological testing with an enzyme-linked immunosorbent assay. Once infected there is no specific treatment.

Infected mosquitoes can give the disease to people and animals. It is common in parts of Africa. People have also contracted the virus in Saudi Arabia and Yemen [115,116].

In humans, the virus can cause several syndromes. Usually, sufferers have either no symptoms or only a mild illness with fever, headache, muscle pains, and liver abnormalities. In a small percentage of cases (<2%), the illness can progress to hemorrhagic fever syndrome, meningoencephalitis (inflammation of the brain and tissues lining the brain), or affect the eye. Patients who become ill usually experience fever, generalized weakness, back pain, dizziness, and weight loss at the onset of the illness. Typically, people recover within two to 7 days after onset. About 1% of people with the disease die of the disease. In livestock, the fatality level is significantly higher. Other signs in livestock include vomiting and diarrhea, respiratory disease, fever, lethargy, anorexia, and sudden death in young animals [117–119].

There is a human vaccine; but, as of 2010, it is not widely available. There is no specific treatment and medical efforts are supportive [118,120].

## Kunjin virus

The main mosquito associated with the spread of Kunjin virus is *Culex annulirostris,* which is geographically widespread and is associated with freshwater habitats. The kunjin virus is a zoonotic virus from the family Flaviviridae and the genus *Flavivirus*. The virus was isolated in mosquitoes in South East Asia but in humans, only in Australia [121,122]. Symptoms of Kunjin virus disease vary. The vast majority of infected people do not develop any symptoms. A small number of people may experience a mild illness with symptoms including fever, headache, muscle pain, swollen lymph nodes, fatigue, and rash. Some of those people may experience encephalitis. Symptoms of encephalitis may include confusion, drowsiness, and seizures. There is no specific treatment for the disease [121,123,124].

## Ross river virus

Ross River virus is endemic to Australia, Papua New Guinea, and other islands in the South Pacific [125]. Ross River virus is transmitted from animals to humans by a number of different types of mosquitoes with *Culex annulirostris, Aedes vigilax* (salt marsh mosquito) and *Aedes notoscriptus* being most common. Ross River virus causes epidemic polyarthritis. The symptoms may include fever with joint pain and swelling which may then be followed in 1 to 10 days by a raised red rash affecting mainly the trunk and limbs. The rash usually lasts for 1 to 10 days and may or may not be accompanied by a fever. The joint pain can be severe and usually lasts 2 to 6 weeks [126–128].

## Barmah forest virus

Barmah Forest virus is an *Alphavirus*. This disease was named after the area in Northern Victoria where it was first isolated in 1974. As of 2015, it has been found only in Australia [129–131]. Barmah Forest virus causes inflammation and joint pain and has similar symptoms to Ross River virus infection (epidemic polyarthritis), but usually, lasts for a shorter duration. The symptoms may include fever, headache, tiredness, painful joints, joint swelling, muscle tenderness, and skin rashes. Some people, especially children, may become infected without showing any symptoms. The initial fever and discomfort only last a few days but some people may experience joint pain, tiredness, and muscle tenderness for up to 6 months. There is no specific drug treatment for Barmah Forest virus infection [130,132–134].

## Jamestown canyon virus

Jamestown Canyon virus, which can be transmitted to several different species of mosquitoes throughout Minnesota, is a rarely reported cause of illness in humans [135]. The virus is closely related to La Crosse virus, although the disease is reported less frequently and any age group may be affected.

About 2 days to 2 weeks after the bite from an infected mosquito, disease symptoms are nonspecific summertime illness with a sore throat, runny nose, and cough, followed by fever, headache, nausea, and vomiting may develop. The neuroinvasive disease occurs in two-thirds of reported cases and is characterized by a severe headache and neck stiffness as in meningitis or increasing lethargy and altered mental status up to coma as in meningoencephalitis. No specific therapy exists for arboviral infections; treatment is limited to supportive care and managing complications, such as relieving increased intracranial pressure [136].

## Eastern equine encephalitis virus

Eastern equine encephalitis (EEE) is a rare illness in humans, and only a few cases are reported in the United States each year. It is among the most serious of a group of mosquito-borne arboviruses that can affect the central nervous system and cause severe complications and even death. EEE is found in freshwater hardwood swampland in the Atlantic and Gulf Coast states in the eastern part of North America, Central and South America, and the Caribbean. It has a complex life cycle involving birds and a specific type of mosquitoes including several *Culex* species and *Culiseta melanura* [137].

Symptoms can range from none at all to a mild flu-like illness fever, headache, and sore throat. More serious infections of the central nervous system lead to a sudden fever and severe headache followed quickly by seizures and coma. About half of these patients die from the disease. Of those who survive, many suffer permanent brain damage and require lifetime institutional care. There is no specific treatment. A vaccine is available for horses, but not humans [51,52,138].

## Zika virus disease

Zika virus is a mosquito-borne flavivirus, which was first isolated in Africa in 1947 [139]. The first documented outbreak among people occurred in 2007 in the Federated States of Micronesia [140]. An outbreak started in Brazil in 2015, and spread to the Americas, Pacific, Asia, and Africa [141]. This led to the World Health Organization declaring it a Public Health Emergency of International Concern in February 2016 [141]. The emergency was lifted in November 2016, but 84 countries still reported cases as of March 2017 [142].

The most common signs and symptoms of zika fever are fever, rash, conjunctivitis, muscle and joint pain, and headache [143]. The time from a mosquito bite to developing symptoms is a few days to a week [144]. The disease lasts for several days to a week and is usually mild enough that people do not have to go to a hospital [145].

Zika virus can be identified by reverse transcriptase PCR (RT-PCR) in acutely ill patients. However, the period of viremia can be short [146] and the World Health Organization (WHO) recommends RT-PCR testing be done on serum collected within 1–3 days of symptom onset or on saliva samples collected during the first 3–5 days [147]. Later on, serology for the detection of specific IgM and IgG antibodies to Zika virus can be used. IgM antibodies can be detectable within 3 days of the onset of illness [148].

There is currently no specific treatment for zika virus infection. Care is supportive with treatment of pain, fever, and itching [147]. Zika virus had been relatively little studied until the major outbreak in 2015, and no specific antiviral treatments are available as yet [145]. Advice to pregnant women is to avoid any risk of infection so far as possible due to the risk of fetal malformation. Once infected there is little treatment that can be provided beyond supportive treatment [149].

## References


[1] Mosquito life cycle. United States environmental protection agency. Assessed at: https://www.epa.gov/mosquitocontrol/mosquito-life-cycle. Last accessed: Nov 9, 2018.

[2] C. Torgan. How mosquitoes detect people. Accessed at: https://www.nih.gov/news-events/nih-research-matters/how-mosquitoes-detect-people. Last accessed: Nov 19, 2018.

[3] Vinauger C, Lutz EK, Riffell JA. Olfactory learning and memory in the disease vector mosquito *Aedes aegypti*. J Exp Biol 2014;217(Pt 13):2321–30.

[4] Success in mosquito control: an integrated approach. United States environmental protection agency. Accessed at: https://www.epa.gov/mosquitocontrol/success-mosquito-control-integrated-approach. Last accessed: Nov 29, 2018.

[5] Controlling mosquitoes at home. Centers for disease control and prevention. Assessed at: https://www.cdc.gov/zika/prevention/controlling-mosquitoes-at-home.html Last assessed: Nov 29, 2018.

[6] Elliott R. Larvicidal control of peridomestic mosquitoes. Trans R Soc Trop Med Hyg November 1, 1955;49(Issue 6):528–42.

[7] Controlling Adult Mosquitoes. United States Environmental Protection Agency. Accessed at: https://www.epa.gov/mosquitocontrol/controlling-adult-mosquitoes. Last accessed: Dec 7, 2018.

[8] Control. The American Mosquito Control Association. Accessed at: https://www.mosquito.org/page/control. Last accessed: Dec 9, 2018.

[9] Brathwaite Dick O, San Martín JL, Montoya RH, del Diego J, Zambrano B, Dayan GH. The history of dengue outbreaks in the Americas. Am J Trop Med Hyg 2012;87:584–93.

[10] Yalcindag E, Elguero E, Arnathau C, Durand P, Akiana J, Anderson TJ, et al. Multiple independent introductions of *Plasmodium falciparum* in South America. Proc Natl Acad Sci USA 2012;109:511–6.

[11] Curtis RB. A history of mosquitoes in Massachusetts.

[12] Harden FW, Hepburn HR, Ethridge B. A history of mosquitoes and mosquito-borne diseases in Mississippi 1699–1965. Mosq News 1967;27:60–6.

[13] Kakkilaya BS. Malaria in wars and victims. http://www.malariasite.com/tag/civil-war.

[14] Jobin W. How did USA get rid of malaria?. https://malariaworld.org/blog/how-did-usa-get-rid-malaria.

[15] Humphreys M. Water won't run uphill: the new deal and malaria control in the American South, 1933–1940. Parassitologia 1998;40:183–91.

[16] Arnold JN, editor. Narragansett historical register: early owners of land in West Greenwich, RI. Hamilton (Rhode Island). Narragansett Historical Publishing Company; 1884. 136–8, https://archive.org/details/narragansetthistv3rhod.

[17] Crosby MC. The American plague: the untold story of yellow fever, the epidemic that shaped our history. New York: Penguin; 2007.

[18] Patterson KD. Yellow fever epidemics and mortality in the United States, 1693–1905. Soc Sci Med 1992;34:855–65.

[19] McCullough D. The path between the seas: the creation of the Panama Canal, 1870–1914. New York: Simon and Schuster; 2001.

[20] Pinheiro FP, Corber SJ. Global situation of dengue and dengue haemorrhagic fever, and its emergence in the Americas. World Health Stat Q 1997;50:161–9.

[21] Centers for Disease Control and Prevention. Elimination of malaria in the United States (1947–1951).

[22] Filler SJ, MacArthur JR, Parise M, Wirtz R, Eliades MJ, Dasilva A, et al. Centers for disease control and prevention. Locally acquired mosquito-transmitted malaria: a guide for investigations in the United States. MMWR Recomm Rep 2006;55(RR-13):1–9.

[23] Centers for Disease Control and Prevention. Eastern equine encephalitis. https://www.cdc.gov/EasternEquineEncephalitis.

[24] Centers for Disease Control and Prevention. Zika cases in the United States. Cumulative Zika virus diseases case counts in the United States, 2015–2017. https://www.cdc.gov/zika/reporting/case-counts.html.

[25] Murray KO, Ruktanonchai D, Hesalroad D, Fonken E, Nolan MS, November. West Nile virus, Texas, USA, 2012. Emerg Infect Dis 2013;19(11):1836–8.

[26] Asnis DS, Conetta R, Teixeira AA, Waldman G, Sampson BA. The West Nile virus outbreak of 1999 in New York: the flushing hospital experience. Clin Infect Dis March 2000;30(3):413–8.

[27] Calisher CH. West Nile virus in the New World: appearance, persistence, and adaptation to a new econiche—an opportunity taken. Viral Immunol 2000;13(4):411–4.

[28] Chen C, Jenkins E, Epp T, Waldner C, Curry PS, Soos C. Climate change and West Ndile virus in a highly endemic region of North America. Int J Environ Res Public Health 2013-07-22;10(7):3052–71.

[29] Hayes EB, Komar N, Nasci RS, Montgomery SP, O'Leary DR, Campbell GL. Epidemiology and transmission dynamics of West Nile virus disease. Emerging Infect. Dis. 2005;11(8):1167–73.

[30] Nash D, Mostashari F, Fine A, et al. The outbreak of West Nile virus infection in the New York City area in 1999. N Engl J Med June 2001;344(24):1807–14.

[31] Anninger WV, Lomeo MD, Dingle J, Epstein AD, Lubow M. West Nile virus-associated optic neuritis and chorioretinitis. Am J Ophthalmol 2003;136(6):1183–5.

[32] Carson PJ, Konewko P, Wold KS, et al. Long-term clinical and neuropsychological outcomes of West Nile virus infection. Clin Infect Dis 2006;43(6):723–30.

[33] Davis LE, DeBiasi R, Goade DE, , et alSep. West Nile virus neuroinvasive disease. Ann Neurol 2006;60(3):286–300.

[34] Montgomery SP, Chow CC, Smith SW, Marfin AA, O'Leary DR, Campbell GL. Rhabdomyolysis in patients with west nile encephalitis and meningitis. Vector Borne Zoonotic Dis 2005;5(3):252–7.

[35] Papa A, Karabaxoglou D, Kansouzidou A, October. Acute West Nile virus neuroinvasive infections: cross-reactivity with dengue virus and tick-borne encephalitis virus. J Med Virol 2011;83(10):1861–5.

[36] Smith RD, Konoplev S, DeCourten-Myers G, Brown T. West Nile virus encephalitis with myositis and orchitis. Hum Pathol February 2004;35(2):254–8.

[37] Mojumder DK, Agosto M, Wilms H, , et alMarch. Is initial preservation of deep tendon reflexes in West Nile Virus paralysis a good prognostic sign? Neurol Asia 2014;19(1):93–7.

[38] Paddock CD, Nicholson WL, Bhatnagar J, et al. Fatal hemorrhagic fever caused by West Nile virus in the United States. Clin Infect Dis June 2006;42(11):1527–35.

[39] Shaikh S, Trese MT. West Nile virus chorioretinitis. Br J Ophthalmol 2004;88(12). 1599–60.

[40] Fonseca K, Prince GD, Bratvold J,, et al. West Nile virus infection and conjunctive exposure. Emerging Infect Dis 2005;11(10):1648–9.
[41] Klee AL, Maidin B, Edwin B, et al. Long-term prognosis for clinical West Nile virus infection. Emerg Infect Dis August 2004;10(8):1405–11.
[42] Watson JT, Pertel PE, Jones RC, , et alSeptember. Clinical characteristics and functional outcomes of West nile fever. Ann Intern Med 2004;141(5):360–5.
[43] Caglioti C, Lalle E, Castilletti C, Carletti F, Capobianchi MR, Bordi L. Chikungunya virus infection: an overview. New Microbiol July 2013;36(3):211–27.
[44] Fourie ED, Morrison JG. Rheumatoid arthritic syndrome after chikungunya fever. S Afr Med J July, 1979;56(4):130–2.
[45] MacFadden DR, Bogoch II. Chikungunya. Can Med Assoc J 2014;186(10). 775–775.
[46] Mahendradas P, Ranganna SK, Shetty R, Balu R, Narayana KM, Babu RB, Shetty BK. Ocular manifestations associated with chikungunya. Ophthalmology February 2008;115(2):287–91.
[47] Powers A. Chikungunya. CDC. Retrieved 12 May 2014.
[48] Schilte C, Staikowsky F, Staikovsky F, Couderc T, Madec Y, Carpentier F, Kassab S, Albert ML, Lecuit M, Michault A. Chikungunya virus-associated long-term arthralgia: a 36-month prospective longitudinal study. PLoS Neglected Trop Dis 2013;7(3):e2137.
[49] Weaver SC, Lecuit M. Chikungunya virus and the global spread of a mosquito-borne disease. N Engl J Med 2015;372(13):1231–9.
[50] Weaver SC, Osorio JE, Livengood JA, Chen R, Stinchcomb DT. Chikungunya virus and prospects for a vaccine. Expert Rev Vaccines 2012;11(9):1087–101.
[51] Eastern equine encephalitis. CDC; August 16, 2010. Retrieved August 7, 2012.
[52] Eastern equine encephalitis fact sheet. CDC; August 16, 2010. Retrieved August 30, 2015.
[53] Campbell GL, Hills SL, Fischer M, Jacobson JA, Hoke CH, Hombach JM, Marfin AA, Solomon T, Tsai TF, Tsu VD, Ginsburg AS, November. Estimated global incidence of Japanese encephalitis: a systematic review. Bull World Health Organ 2011;89(10):766–74.
[54] Ghosh D, Basu A. Japanese encephalitis-a pathological and clinical perspectiveBrooker S, editor. PLoS Negl Trop Dis September 2009;3(9):e437.
[55] Kim HC, Terry A, Klein, RatreeTakhampunya, Evans BP, SirimaMingmongkolchai, AmpornpanKengluecha, Grieco J, Masuoka P, Kim M-S, Chong S-T, Jong-Koo Lee, Lee W-J. Japanese encephalitis virus in culicine mosquitoes (Diptera: Culicidae) collected at Daeseongdong, a village in the demilitarized zone of the Republic of Korea. J Med Entomol 2011;48(6):1250–6.
[56] Ghoshal A, Das S, Ghosh S, Mishra MK, Sharma V, Koli P, Sen E, Basu A. Proinflammatory mediators released by activated microglia induces neuronal death in Japanese encephalitis. Glia 2007;55(5):483–96.
[57] Jelinek T. Japanese encephalitis vaccine in travelers. Expert Rev Vaccines July 2008;7(5):689–93.
[58] Gupta N, Lomash Vinay, Rao PV. Expression profile of Japanese encephalitis virus induced neuroinflammation and its implication in disease severity. J Clin Virol September 2010;49(1):04–10.
[59] Solomon T, Dung NM, Kneen R, Gainsborough M, Vaughn DW, Khanh VT. Japanese encephalitis. J Neurol Neurosurg Psychiatry 2000;68(9):405–15.
[60] Su HL, Liao CL, Lin YL. Japanese encephalitis virus infection initiates endoplasmic reticulum stress and an unfolded protein response. J Virol May 2002;76(9):4162–71.
[61] Schiøler KL, Samuel M, Wai KL. Vaccines for preventing Japanese encephalitis. Cochrane Database Syst Rev 2007;3:CD004263.

[62] Murray Valley encephalitis (MVE) factsheet. New South Wales Department of Health; March 2, 2011. Retrieved 30 December 2011.

[63] Burnet FM. Murray Valley encephalitis. Am J Public Health Nation's Health 1952;42(12):1519–21.

[64] French EL. Murray Valley encephalitis isolation and characterization of the aetiological agent. Med J Aust 1952;1(4):100–3.

[65] Hurrelbrink RJ, Nestorowicz A, McMinn PC. Characterization of infectious Murray Valley encephalitis virus derived from a stably cloned genome-length cDNA. J Gen Virol December, 1999;80(12):3115–25.

[66] Thompson WH, Kalfayan B, Anslow RO. Isolation of California encephalitis virus from a fatal human illness. Am J Epidemiol 1965;81:245–53.

[67] La Crosse virus disease cases and deaths reported to CDC by year and clinical presentation, 2004–2013(PDF). Centers for Disease Control and Prevention. Retrieved 4 December 2016.

[68] Center for Disease Control and Prevention (CDC). Possible congenital infection with La crosse encephalitis virus – West Virginia, 2006–2007. MMWR Morb Mortal Wkly Rep January 2009;58(1):4–7.

[69] McJunkin JE. Safety and pharmacokinetics of ribavirin for the treatment of la crosse encephalitis. Pediatr Infect Dis J October 2011;30(10):860–5.

[70] McJunkin JE, de los Reyes EC, Irazuzta JE, Caceres MJ, Khan RR, Minnich LL, Fu KD, Lovett GD, Tsai T, Thompson A. La crosse encephalitis in children (pdf). N Engl J Med March 2001;344(11):801–7.

[71] Bhatt S, Weiss JD, Cameron E, Bisanzio D, Mappin B, Dalrymple U, Battle KE, Moyes CL, Henry A, Eckhoff PA, Wenger EA, Briët O, Penny MA, Smith TA, Bennett A, Yukich J, Eisele TP, Griffin JT, Fergus AC, Lynch M, Lindgren F, Cohen JM, Murray CLJ, Smith DL, Hay SI, Cibulskis RE, Gething PW. The effect of malaria control on Plasmodium falciparum in Africa between 2000 and 2015. Nature September 16 , 2015;526(7572):207–11.

[72] Korenromp E, Williams B, de Vlas S, Gouws E, Gilks C, Ghys P, Nahlen B. Malaria attributable to the HIV-1 epidemic, sub-Saharan Africa. Emerg Infect Dis 2005;11(9):1410–9.

[73] Layne SP. Principles of infectious disease Epidemiology(PDF). EPI 220. UCLA Department of Epidemiology. Retrieved 2007-06-15.

[74] Murray CJ, Rosenfeld LC, Lim SS, Andrews KG, Foreman KJ, Haring D, Fullman N, Naghavi M, Lozano R, Lopez AD. Global malaria mortality between 1980 and 2010: a systematic analysis. Lancet 2012;379(9814):413–31.

[75] Olupot-Olupot P, Maitland K. Management of severe malaria: results from recent trials. Advances in experimental medicine and biology. Adv Exp Med Biol 2013;764:241–50.

[76] World malaria report 2015. World Health Organization; December 2015.

[77] Cowman AF, Berry D, Baum J. The cellular and molecular basis for malaria parasite invasion of the human red blood cell. J Cell Biol 2012;198(6):961–71.

[78] Tilley L, Dixon MW, Kirk K. The Plasmodium falciparum-infected red blood cell. Int J Biochem Cell Biol 2011;43(6):839–42.

[79] Tran TM, Samal B, Kirkness E, Crompton PD. Systems immunology of human malaria. Trends Parasitol 2012;28(6):248–57.

[80] Vaughan AM, Aly AS, Kappe SH. Malaria parasite pre-erythrocytic stage infection: gliding and hiding. Cell Host Microbe 2008;4(3):209–18.

[81] Beare NA, Taylor TE, Harding SP, Lewallen S, Molyneux ME. Malarial retinopathy: a newly established diagnostic sign in severe malaria. Am J Trop Med Hyg 2006;75(5):790–7.

[82] Caraballo H. Emergency department management of mosquito-borne illness: malaria, dengue, and west nile virus. Emerg Med Pract 2014;16(5).

[83] Greenwood BM, Bojang K, Whitty CJ, Targett GA. Malaria. Lancet 2005;365(9469):1487–98.

[84] Koella JC, Sorensen, May A(7. The malaria parasite, Plasmodium falciparum, increases the frequency of multiple feeding of its mosquito vector, *Anopheles gambiae*. Proc Biol Sci 1998;265(1398):763–8.

[85] Lalloo DG, Olukoya P, Olliaro P. Malaria in adolescence: burden of disease, consequences, and opportunities for intervention. Lancet Infect Dis 2006;6(12):780–93.

[86] Nadjm B, Behrens RH. Malaria: an update for physicians. Infect Dis Clin N Am 2012;26(2):243–59.

[87] Organization, World Health. Guidelines for the treatment of malaria. 2nd ed. Geneva: World Health Organization; 2010.

[88] White NJ. Determinants of relapse periodicity in Plasmodium vivax malaria. Malar J 2011;10:297.

[89] WHO. World malaria report 2014. Geneva, Switzerland: World Health Organization; 2014. p. 32–42.

[90] Baird JK. Evidence and implications of mortality associated with acute plasmodium vivax malaria. Clin Microbiol Rev 2013;26(1):36–57.

[91] Idro R, Marsh K, John CC, Newton CR, October. Cerebral malaria: mechanisms of brain injury and strategies for improved neurocognitive outcome. Pediatr Res 2010;68(4):267–74.

[92] Kochar DK, Saxena V, Singh N, Kochar SK, Kumar SV, Das A. Plasmodium vivax malaria. Emerg Infect Dis January 2005;11(1):132–4.

[93] Rénia L, Wu Howland S, Claser C, Charlotte Gruner A, Suwanarusk R, HuiTeo T, Russell B, Ng LF. Cerebral malaria: mysteries at the blood-brain barrier. Virulence 2012;3(2):193–201.

[94] Sarkar PK, Ahluwalia G, Vijayan VK, Talwar A. Critical care aspects of malaria. J Intensive Care Med 2009;25(2):93–103.

[95] Trampuz A, Jereb M, Muzlovic I, Prabhu R. Clinical review: severe malaria. Crit Care 2003;7(4):315–23.

[96] Achan J, Talisuna AO, Erhart A, Yeka A, Tibenderana JK, Baliraine FN, Rosenthal PJ, D'Alessandro U. Quinine, an old anti-malarial drug in a modern world: role in the treatment of malaria. Malar J 2011;10(1):144.

[97] Kajfasz P. Malaria prevention. Int Marit Health 2009;60(1–2):67–70.

[98] Kokwaro G. Ongoing challenges in the management of malaria. Malar J 2009;8(Suppl. 1):S2.

[99] Waters NC, Edstein MD. 8-Aminoquinolines: primaquine and tafenoquine. In: Staines HM, Krishna S, editors. Treatment and prevention of malaria: antimalarial drug chemistry, action and use. Springer; 2012. p. 69–93.

[100] Pasvol G. The treatment of complicated and severe malaria. Br Med Bull 2005;75–76:29–47.

[101] Sinha S, Medhi B, Sehgal R. Challenges of drug-resistant malaria. Parasite 2014;21:61.

[102] Auguste AJ, Pybus OG, Carrington CV. Evolution and dispersal of St. Louis encephalitis virus in the Americas. Infect Genet Evol 2009;9(4):709–15. Baillie GJ, Kolokotronis SO,

Waltari E, Maffei JG, Kramer LD, Perkins SL. (2008) Phylogenetic and evolutionary analyses of St. Louis encephalitis virus genomes. Molecular Phylogenetics and Evolution 47(2):717-728.

[103] Bredeck JF, November. The story of the epidemic of encephalitis in St. Louis. Am J Public Health Nation's Health 1933;23(11):1135–40.

[104] Kramer LD, Chandler LJ. Phylogenetic analysis of the envelope gene of St. Louis encephalitis virus. Arch Virol 2001;146(12):2341–55.

[105] May FJ, Li L, Zhang S, Guzman H, Beasley DW, Tesh RB, Higgs S, Raj P, Bueno Jr R, Randle Y, Chandler L, Barrett AD. Genetic variation of St. Louis encephalitis virus. J Gen Virol 2008;89(8):1901–10.

[106] Encephalitis. In St. Louis. Am J Public Health Nation's Health October 1933;23(10):1058–60.

[107] Rahal JJ, Anderson J, Rosenberg C, Reagan T, Thompson LL. Effect of interferon-alpha2b therapy on St. Louis viral meningoencephalitis: clinical and laboratory results of a pilot study. J Infect Dis 2004;15;190(6):1084–7.

[108] Barrett AD, Higgs S. Yellow fever: a disease that has yet to be conquered. Annu Rev Entomol 2007;52:209–29.

[109] Heymann, D., ed. 2008. Control of Auguste AJ, Lemey P, Pybus OG, suchard MA, Salas RA, Adesiyun AA, Barrett AD, Tesh RB, Weaver SC, Carrington CV. Yellow fever virus maintenance in Trinidad and its dispersal throughout the Americas. J Virol 2010;84(19):9967–77.

[110] Barrett AD, Higgs S. Yellow fever: a disease that has yet to be conquered. Annu Rev Entomol 2007;vol. 52:209–29. Communicable diseases manual, 19th ed.. Washington, DC: American Public Health Association.

[111] Frequently asked questions about yellow fever. CDC; August 21, 2015. Retrieved 18 March 2016.

[112] Rogers DJ, Wilson AJ, Hay SI, Graham AJ. The global distribution of yellow fever and dengue. Adv Parasitol 2006;62:181–220.

[113] Staples JE, Monath TP. Yellow fever: 100 years of discovery. J Am Med Assoc August 27, 2008;300(8):960–2.

[114] Australian Government. The Australian immunisation handbook. 10th ed. 2013.

[115] Arzt J, White WR, Thomsen BV, Brown CC. Agricultural diseases on the move early in the third millennium. Vet Pathol January 2010;47(1):15–27.

[116] Bird BH, Ksiazek TG, Nichol ST, Maclachlan NJ. Rift Valley fever virus. J Am Vet Med Assoc April 2009;234(7):883–93.

[117] Boshra H, Lorenzo G, Busquets N, Brun A. Rift Valley fever: recent insights into pathogenesis and prevention. J Virol 2011-07-01;85(13):6098–105.

[118] Rift Valley fever. Fact sheet N°207. World Health Organization. May 2010. Retrieved 21 March 2014.

[119] Swanepoel R, Coetzer JA. Rift Valley fever. In: Coetzer JA, Tustin RC, editors. Infectious diseases of livestock. 2nd ed. CapeTown: OxfordUniversityPressSouthernAfrica; 2004. p. 1037–70.

[120] Bird BH, Maartens LH, Campbell S, Erasmus BJ, Erickson BR, Dodd KA, Spiropoulou CF, Cannon D, Drew CP. Rift Valley fever virus vaccine lacking the NSs and NSm genes is safe, nonteratogenic, and confers protection from viremia, pyrexia, and abortion following challenge in adult and pregnant sheep. J Virol 2011-12-01;85(24):12901–9.

[121] Cook GC, Zumla AI. Manson's tropical diseases. Elsevier Health Sciences; 2008. p. 736–7.

[122] Mackenzie JS, Ashford RW, Service MW. Encyclopedia of arthropod-transmitted infections of man and domesticated animals. CABI; 2001. p. 251.

[123] Department of health and ageing—kunjin virus infection—fact sheet. Government of Australia; 2004. Retrieved 2009-08-08.

[124] Hirsh DC, Nigel James Maclachlan, Walker RL. Veterinary microbiology. Wiley-Blackwell; 2004. p. 354.

[125] Harley D, Sleigh A, Ritchie S. Ross River virus transmission, infection, and disease: a cross-disciplinary review. Clin Microbiol Rev 2001;14(4):909–32 [tableofcontents].

[126] Fraser JR. Epidemic polyarthritis and Ross River virus disease. Clin Rheum Dis August 12, 1986;12:369–88.

[127] Ross River virus disease. Better health channel. Department of Health, State Government, Victoria. March 2013. Retrieved 15 April 2015.

[128] Schleenvoigt BT, Baier M, Hagel S, Forstner C, Kotsche R, Pletz MW. Ross river virus infection in a Thuringian traveller returning from south-east Australia. Infection 2015;43(2):229–30.

[129] Cashman P. Barmah Forest virus serology: implications for diagnosis and public health action. Comm Dis Intell 2008-06-30. Retrieved 2015-04-07.

[130] Kostyuchenko VA, Jakana J, Liu X, Haddow AD, Aung M, Weaver SC, Chiu W, Lok S-M. The structure of Barmah forest virus as revealed by cryo-electron microscopy at a 6-Angstrom resolution has detailed transmembrane protein architecture and interactions. J Virol 2011;85(18):9327–33.

[131] Smith DW, May. The viruses of Australia and the risk to tourists. Travel Med Infect Di 2011;9:113–25.

[132] Ehlkes, Lutz. Surveillance should be strengthened to improve epidemiological understandings of mosquito-borne Barmah Forest virus infection. Western Pac Surveill Response J September 2012;3:63–8.

[133] Naish S, Mengersen K, Hu W, Tong S. Forecasting the future risk of Barmah Forest virus disease under climate change scenarios in Queensland, Australia. PLoS One 2013;8(5):E62843.

[134] Naish S, Hu W, Mengersen K, Tong Shilu. Spatio-temporal patterns of Barmah forest virus disease in Queensland, Australia. PLoS One October 13, 2011;6(10):e25688.

[135] Pastula DM, Hoang Johnson DK, White JL, Dupuis 2nd AP, Fischer M, Staples JE, August 5. Jamestown Canyon virus disease in the United States—2000–2013. Am J Trop Med Hyg 2015;93(2):384–9.

[136] Lindsey NP, Lehman JA, Staples JE, Fischer M. Division of vector-borne diseases, National center for emerging and zoonotic infectious diseases, CDC. West nile virus and other arboviral diseases–United States, 2013. Morbid Mortal Wkly Rep CDC June 20, 2014;63(24):521–6.

[137] Zheng Y. Mosquito-borne virus infects 2d in Mass. The Boston Globe; August 16, 2008.

[138] Arrigo NC, Adams AP, Weaver SC. Evolutionary patterns of eastern equine encephalitis virus in North versus South America suggest ecological differences and taxonomic revision. J Virol 2010;84(2):1014–25.

[139] Olson KE, Haddow AD, Schuh AJ, et al. Genetic characterization of zika virus strains: geographic expansion of the Asian lineage. PLoS Neglected Trop Dis 2012;6(2):e1477. https://doi.org/10.1371/journal.pntd.0001477. ISSN 1935-2735. *PMC* 3289602. *PMID* 22389730.

[140] Chen LH, Hamer DH. Ideas and opinions. Zika virus: rapid spread in the western hemisphere. Ann Intern Med 2016;164(9):613–5.

[141] WHO Director-General summarizes the outcome of the Emergency Committee regarding clusters of microcephaly and Guillain–Barré syndrome. *Media Centre*. World Health Organization; February, 2016. Archived from the original on 3 February 2016. Retrieved 3 February 2016.

[142] Zika situation report". World health organization. Archived from the original on 25 may 2017. Retrieved 26 May 2017.

[143] Heang V, Yasuda CY, Sovann L, et al. Zika virus infection, Cambodia. Emerg Infect Dis 2012;18(2):349–51.

[144] Signs and symptoms. Zika virus home. Centers for disease control and prevention. November 5, 2014. Archived from the original on 29 January 2016. Retrieved 30 January 2016.

[145] Sikka V, Chattu VK, Popli RK, et al. The emergence of zika virus as a global health security threat: a review and a consensus statement of the INDUSEM Joint working Group (JWG)". J Glob Infect Dis February 11, 2016;8(1):3–15.

[146] Factsheet for health professionals". Zika virus infection. European Centre for Disease Prevention and Control. Archived from the original on 23 December 2015. Retrieved 22 December 2015.

[147] Zika virus". Archived *from the original on 25 December 2015*. Retrieved *24 December 2015*.

[148] Hayes EB. Zika virus outside Africa. Emerg Infect Dis 2009;15(9):1347–50.

[149] Petersen EE, Staples JE, Meaney-Delman D, et al. Interim guidelines for pregnant women during a zika virus outbreak — United States, 2016. MMWR (Morb Mortal Wkly Rep) 2016;65(2):30–3.

# Chapter 6

# Viral genetics and structure

**Iqra Naveed Akhtar**

*Clinical Research Fellow, Department of Neurology, University of Missouri, Columbia, MO, United States; Zeenat Qureshi Stroke Institute, St. Cloud, MN, United States*

## Genomic organization of the Dengue virus: an introduction

The viral genome of the Dengue virus is a linear, single-stranded, positive sense RNA molecule ($\approx$11 kb) translated as a single open reading frame (ORF), bordered by 5′ and 3′ untranslated regions (UTRs) on each side [1] [summarized in Fig. 6.1]. Through the combination of cryoelectron microscopy, X-ray crystallography, and imaging reconstruction, the structure of the Dengue virus particle was deciphered as a spherical 50 nm virion [2]. Multiple copies of the capsid (C) protein encapsulates the ssRNA genome, creating the

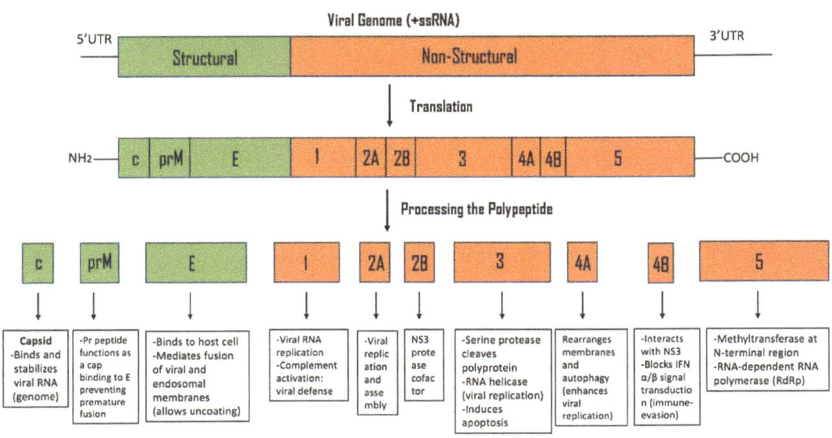


**FIGURE 6.1 An Illustration of the Dengue Viral Genome.** Brief functions of each Structural and Nonstructural protein of the Dengue Virus are mentioned in the outlined box below the protein. *Illustration adapted from two combined resources*: *Guzman, M. G et al. Dengue: A continuing global threat.* Nature Reviews Microbiology *(2010) and Khetarpal N et al. Dengue Fever: Causes, Complications, and Vaccine Strategies.* J Immunol Res. *(2016)*

**Dengue Virus Disease. https://doi.org/10.1016/B978-0-12-818270-3.00006-0**

nucleocapsid, which forms the electron-dense core of the Dengue virus particle. The core is further covered by a lipid bilayer into which two transmembrane proteins are positioned, forming an outer glycoprotein protective casing [3]. This protective armor is composed of 180 copies of envelope (E) and membrane (prM/M) proteins, with varying configuration depending on the stage of maturity the virus has met [3]. Upon entering the cell and release from the capsid, the positive ssRNA genome is translated to a polyprotein of 3400 amino acids in length, which is then subsequently cleaved by viral encoded proteases and host proteases to yield 10 proteins: three structural proteins (C, E, prM) and seven nonstructural (NS) proteins, labeled as NS1, NS2A, NS2B, NS3, NS4A, NS4B, and NS5) [1]. The structural proteins are vital in virion assembly, release, maturation, and infectivity; while the nonstructural proteins play a major role in viral replication and in eluding the host's immune response. There are four main serotypes of the Dengue virus (numbered 1–4), each sharing approximately 60%–70% amino acid sequence homology, with a further subdivision of serotypes into genotypes where the similarities begin to wane (at 3%) [4]. While these serotypes can cause wavering presentations in the host, ranging from mild symptoms to life-threatening Dengue hemorrhagic fever, a common feature among all of them, and among all the flaviviruses are the production of these 10 mature proteins, all of which have a unique role to play in the life cycle of Dengue virus.

## Life cycle of Dengue virus: A brief overview

Dengue virus is transmitted by the Aedes mosquito (*Aedes aegypti* or *Aedes albopictus*) and introduced into the host by the infected vector during its blood meal [4]. Flavivirus adheres to target cells, predominately monocytes, macrophages, and dendritic cells, by the envelope protein (E), after which internalization of the virus occurs through clathrin-dependent receptor-mediated endocytosis into endosomes [4]. The acidic environment within the endosome then causes structural rearrangement of the E protein, a necessary feat resulting in fusion between the viral and host cellular membranes. Ubiquitination of the capsid ultimately releases the viral genome into the cytosol [5].

The viral RNA serves as the mRNA template for translation (early response) into a precursor polypeptide, which is then subsequently co- and posttranslated by host and viral proteases into structural and nonstructural proteins. After translation into protein components, an RNA-dependent-RNA polymerase (RdRp) derived from the NS5 protein synthesizes a complementary minus-strand from the positive-strand RNA that makes up the Dengue viral genome [6]. This minus strand then serves as a template for synthesis of a new positive strand that will be used as the Dengue virus genome for future virions. This viral replication process occurs in the cytoplasm. Immature, noninfectious virions then assemble within the endoplasmic reticulum, with

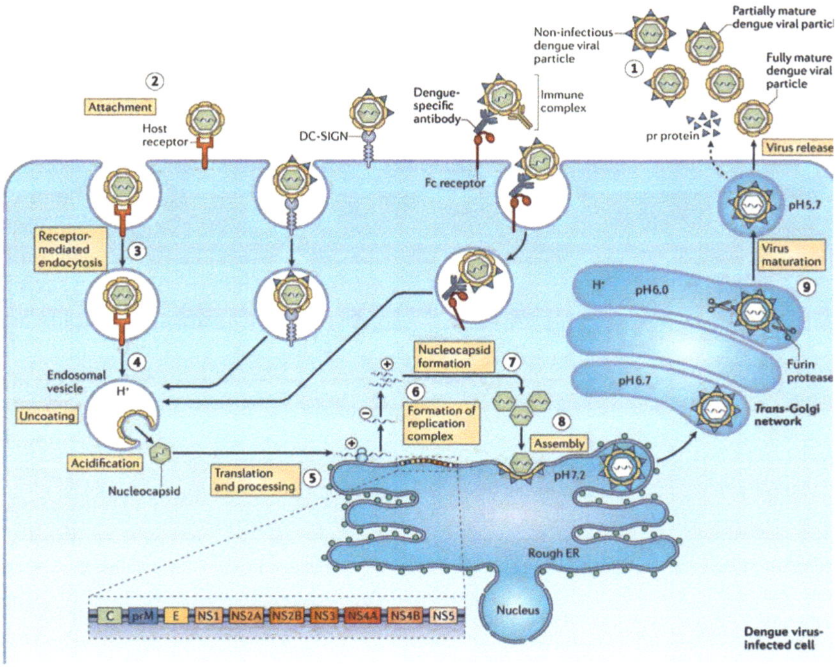


FIGURE 6.2 **Dengue Life Cycle: An Overview.** Dengue virus E protein binds to the target cell (2), initiating receptor-mediated endocytosis (3) of the Dengue virus into endosomes (4). The acidic environment in the endosomes causes uncoating of the virion, with the release of the nucleocapsid (NC) into the cytosol. The viral RNA undergoes an early response, in which a precursor polypeptide is made and cleaved into structural and nonstructural proteins (5). An RNA-dependent RNA polymerase derived from the NS5 protein (nonstructural) initiates viral replication in the cytosol (6). This newly synthesized RNA will then bind to capsid protein in the endoplasmic reticulum, forming a nucleocapsid (7). Assembly of the virion will occur within the rough endoplasmic reticulum (8). The virion then buds off the membrane of the endoplasmic reticulum, acquiring the E protein and prM/M proteins and makes its way to the trans-Golgi network. The acidic environment of the trans-Golgi network will cause a change in structure, yielding a mature Dengue virion (9). Furin cleavage of the prM from the E protein will allow for the release of the mature Dengue virion into the extracellular space, allowing them to infect new cells (10). *Illustration from Screaton, Gavin et al. New insights into the immunopathology and control of dengue virus infection. Nature Reviews Immunology (2015).*

the newly synthesized RNA bound with the capsid protein, forming the nucleocapsid. The virion will then bud off the membrane of the endoplasmic reticulum into the lumen, acquiring the E protein and the prM/M proteins en-route through the lipid membranes [illustrated in Fig. 6.3]. In the trans-Golgi network, the prM will be cleaved by furin, catapulting the release of the mature infectious virions from the cell via exocytosis [6].

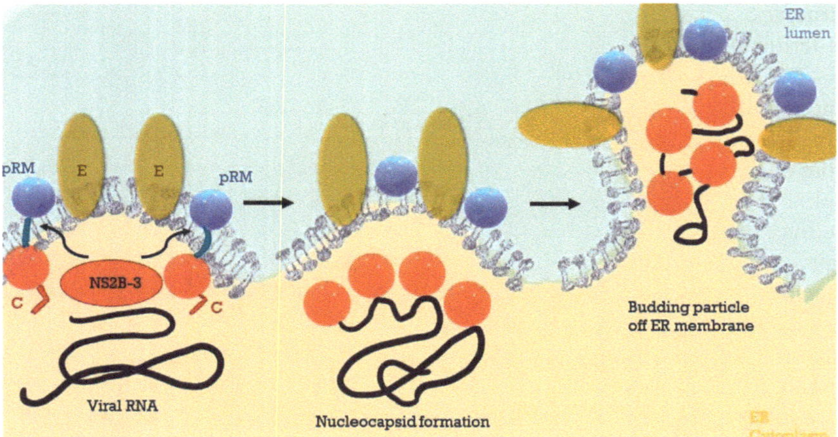


FIGURE 6.3 **Dengue Viral Assembly: A portrayal.** An anchor binds both the C protein and the prM protein to the membrane of the endoplasmic reticulum. The C protein anchor junction is first cleaved by the NS3 protein, which is a serine protease, using NS2b as a cofactor. Occurring right after this cleavage is a second cleavage, at the prM-anchor site performed by a host-encoded signalase. Once the C protein is released from the anchor, it is considered a mature capsid protein, and readily available to bind to dengue viral RNA, forming the nucleocapsid. The Dengue virion buds off the membrane of the endoplasmic reticulum into the lumen acquiring the E protein and the prM/M proteins in the process. *NS2B-3: Nonstructural protein 2B-3; prM: pre-membrane; M: membrane; E: envelope; C: capsid; ER: endoplasmic reticulum. Illustration adapted from Byk LA et al. Properties and functions of the dengue virus capsid protein.* Annu Rev Virol. *(2016)*

# The role of structural proteins

## Dengue viral capsid protein (C)

### *Capsid structure*

The C protein is the first polypeptide synthesized during RNA translation of the Dengue virus: a highly basic moiety (rich in arginine and lysine) of 12 kDa [6]. The basic nature of this 100-residue protein (with 26 basic amino acids compared to only three acidic amino acids) is thought to be the reason that it can cooperate so readily with virion RNA to form a nucleocapsid, the core of the mature Dengue virus particle [6]. Newer nuclear magnetic resonance techniques have allowed detailed stereographic visualization of the Dengue viral capsid protein. This homodimer shows that the basic residues amass predominately on one end of the protein, with an apolar surface on the other end. Each monomer has four $\alpha$ helices. The N-terminal domain, lacking an N-terminal hydrophobic signal sequence, is composed highly of basic amino acids, and shows flexibility, and a more disordered structure, especially in solution [7]. This unveils the fact that although the C protein is conserved fairly well structurally among the flaviviruses, it is also yields the most structural variability among the structural proteins, owing to this unique N-terminal region of the protein.

The $\alpha$-helices are situated in the homodimer in the following manner: $\alpha2$ and $\alpha4$ of one monomer are antiparallel to the $\alpha2$ and $\alpha4$ of the other monomer, forming the crux of the dimer surface [illustrated in Fig. 6.4]. The core of the monomer is formed by the $\alpha1$, $\alpha2$, and $\alpha3$ helices, while the longest helix, the $\alpha4$, is predominately basic in amino acids, with the $\alpha4$-$\alpha4'$ interaction among the monomers vital for dimer formation, production of infectious Dengue virus particles, and protein stability [7,8]. In research studying the specifics of the Dengue virus serotype 4 (DENV4) spearheaded in 1997, researchers found a hydrophobic sequence of residues (from 45 to 65 positions). They concluded that this hydrophobic region was highly conserved among flaviviruses and is responsible for the anchoring of the mature Dengue virus capsid protein to the membrane of the endoplasmic reticulum, the cytoplasmic side serving as the site for viral replication [9]. The mature Dengue virus capsid has a distinctive duality to it: it can bind to both the viral RNA *and* the lipid membranes (such as the membrane of the endoplasmic

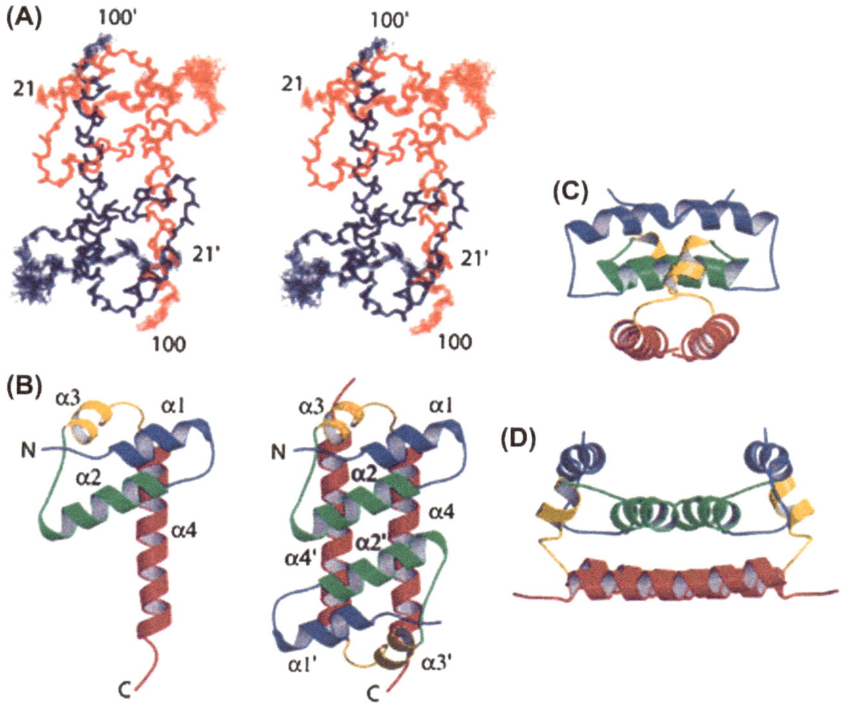


**FIGURE 6.4 The alpha-helical structure of the Dengue virus.** $\alpha1$: blue, $\alpha2$: green, $\alpha3$: yellow, $\alpha4$: red. Illustration by Ma Lixin et al. Solution structure of Dengue virus capsid protein reveals another fold. PNAS (2004). *Copyright (2004) National Academy of Sciences, U.S.A.*

reticulum). This duality can be traced back to the α-helical structure of the homodimer, with the highly basic α4-α4′ interaction interacting with the viral RNA, and the uncharged, apolar α2-α2′ region making up the hydrophobic portion of the capsid, interacting with the membrane of the endoplasmic reticulum [7]. The basic residues composing the N-terminal region are also found to play a part in RNA binding to the capsid protein as well as mature virus particle production.

### Chaperoning role of the capsid

The capsid proteins of Dengue virus can serve as chaperones, safeguarding the viral RNA to which they are bound. Studies investigating the West Nile Virus capsid in vitro are used as a model for this, due to the fact that the Dengue virus capsid aggregates when bound to viral RNA, possibly due to the fact that the positive-sense RNA causes aggregation of the hydrophobic zones of the Dengue virus capsid protein [10]. The capsid protein can prevent misfolding of the viral RNA or even correct misfolded RNA all the while without utilizing ATP. The chaperone activity lies in the key basic domains of the C protein, those of which were shown previously to be highly flexible [11].

### Maturation of the capsid protein

The capsid protein is transiently bound to the membrane of the endoplasmic reticulum on the cytoplasmic side by an "anchor", a hydrophobic signal sequence that spans the membrane, which also binds to the premembrane (prM) protein, a protein that is translated just following the capsid protein, on the luminal side of the endoplasmic reticulum membrane [6]. Cleavage of the capsid protein from the anchor site is a requisite to the maturation of the capsid protein. This cleavage at the capsid-anchor site is accomplished by way of a viral encoded protease, NS3 (using NS2B as a cofactor). At the prM-anchor junction in the lumen of the endoplasmic reticulum, a separate host signal peptidase ("signalase") separates the prM and the anchor from each other (illustrated in Fig. 6.2). It is important that this two-step cleavage at both sites, occurs in this order: with cleavage of the capsid from the anchor occurring first, followed by the cleavage of the prM from the anchor [12,13]. Studies have shown in other flaviviruses, when the order of the sequential cleavage is altered, it can lead to particles being released extracellularly that had no infectious nucleocapsid, while retaining the E and prM proteins on the surface, rendering the "empty" particles [14]. Only when the capsid protein is cleaved from the anchor, is the viral capsid considered mature and ready to bind to viral RNA.

### Capsid binding to RNA: nucleocapsid formation

Assembly of Dengue virus within cells starts with the cardinal step of the nucleocapsid formation. After cleavage by the viral protease NS3, the mature capsid protein is then able to bind to viral RNA especially at the $\alpha4$-$\alpha4'$ helical site on the capsid protein. Viral replication occurs in isolated "vesicle packets", structures derived from endoplasmic reticulum outpouchings and surrounded by membrane. Upon completion of replication, the newly synthesized RNA is then extruded from an opening in the vesicle packets into the cytoplasm, where it can bind to the mature capsid protein [15]. Encapsidation by the viral capsid is mediated by viral encoded signals, of which are poorly understood, and may be a result of immeasurable factors in the noncoding regions of the Dengue viral genome. A proposed prototype showing the transportation of the viral genome to the site of nucleocapsid assembly at the membranes of the endoplasmic reticulum, also illustrates the "buds" that form off the endoplasmic reticulum membrane, acquiring both the E and prM proteins in the process [15–17]. The replication of the RNA genome in these replication pockets (or vesicle packets) is a prerequisite for encapsidation later, as it has been shown that for RNA replicated out of these pockets, encapsidation does not occur [18].

The budding of the Dengue virus nucleocapsid and the acquiring of the E and prM proteins is proposed to both be independent processes, as empty viruses containing surface E and prM but without the nucleocapsid can be produced and extruded from the cell, if the E and prM proteins are overexpressed artificially [19]. The NS3 protease that is responsible for capsid maturation (by cleaving the capsid protein from the anchor) may play a role in nucleocapsid incorporation into buds off the endoplasmic reticulum membrane toward the lumen. This highlights the fact that not only is the NS3 (along with NS2B) required for capsid maturation, but that it may also be significant for nucleocapsid recruitment and viral assembly. However, it is still not known how changes in NS2B or NS3 impair particle formation, whether it is due to binding to the C protein, viral RNA, or other proteins or membranes in the cell. Other nonstructural proteins may also play noteworthy roles in viral assembly, with recent studies pointing toward NS2A and NS1, showing that mutations in these can lead to abhorrent host protein interactions and fewer infectious particles released [20]. Nucleocapsid assembly and the acquiring of the E and prM proteins is a complex process, one that is still not fully uncovered, and may be a result of the interplay between many parts of the viral genome and host proteins, warranting further studies.

## *Mutations or deletions in the capsid*

The complete Dengue viral capsid-coding RNA sequence was reported by a widespread analysis using biochemical probes [21]. The first 160 of the 300 total nucleotides that make up the RNA encoding the capsid, were conserved RNA structures, shared among other viruses in its genome. These *cis*-acting RNA functional units include: (A) RNA cyclization signal (5′CS) vital for RNA synthesis, (B) a hairpin loop (cHP) also essential for RNA synthesis, and (C) a pseudoknot included in the region C1 which augments cyclization of the genome due to its sequence being complementary to a region in the 3′ end of the genome, thus also increasing RNA synthesis [21–23].

Understanding that there are highly conserved sequences in the capsid coding RNA sequence, allows for mutational studies to segregate these conserved *cis*-sequences from purely capsid-coding sequences, to assess the function of various areas of the capsid RNA sequence [24,25]. Alterations in specific portions of the viral capsid sequence were of significance, while others showed no effect on viral RNA synthesis and nucleocapsid assembly. Deletions in the highly basic N-terminal region of the capsid were shown to significantly alter the formation of Dengue viral particles, indicating that they ultimately affect viral assembly. A mutational analysis of this region indicated that at least two positive charges in each of the two main basic clusters in this region (R5K6K7R9 and K17R18R20R22) are necessary, showing that positive charges might play a larger role in viral assembly rather than just the placement of residues in specific positions [24].

In the $\alpha$-helices of the Dengue virus capsid, mutations yielded a wide array of results. Mutations in the $\alpha1$ and the $\alpha1$-$\alpha2$ connecting bend showed no effect on Dengue viral assembly and propagation, while those in the more hydrophobic regions such as the $\alpha2$ helix showed marked suppression of Dengue virus propagation [25]. This importance of the hydrophobic region of the $\alpha2$ helix is a prevailing feature in other flaviviruses, suggesting that even though the capsid structure varies among them, any alterations in the hydrophobic sequences can severely affect nucleocapsid assembly and viral propagation [26–29]. This phenomenon is further reiterated by studies that demonstrated that large deletions or mutations affecting as many as 36 nucleotides was better stomached than smaller mutations/deletions in the N terminal hydrophobic coding sequence [30].

### *Capsid localization*

The Dengue viral capsid is at an ideal location being associated with the membranes of the endoplasmic reticulum: in close vicinity to (A) the vesicle packet exit sites that release newly synthesized viral RNA, and (B) near the budding particles sites that are made by the protrusion of the membrane of the endoplasmic reticulum [15–17,31,32]. While large mutations or deletions involving the capsid protein structure have a negligible effect on viral propagation (in previous section), newer studies have shown that the mutations affecting the localization of the capsid protein are much more consequential, significantly altering viral propagation [25]. There are two main locations where the Dengue viral capsid can be found in infected cells: in the cytoplasm and in the nucleus. In the nucleus, the Dengue viral capsid is found specifically in the nucleolus, while in the cytoplasm it is concentrated both in the membranes of the endoplasmic reticulum and lipid droplet surfaces [6].

## Lipid droplets

Lipid droplets are organelles derived from the ER, with a core of neutral lipids and an outer layer of phospholipids and protein [33]. In a study investigating the subcellular localization of cytoplasmic C protein, BHK cells were infected with the Dengue virus serotype 2. Visualization of the dengue virus particles after fixation by paraformaldehyde and Triton X-100 showed that the C protein gathered in spherical shapes, with the C protein in a ring-like pattern [25]. ER and Golgi markers showed that it was not accumulated in these regions. The ring-like structure of the C protein resembled the structure of hepatitis C (HCV) which is known to accumulate on the surface of the lipid droplets [34–36]. This led researchers to believe that like HCV, the Dengue virus capsid accumulation on the surface of lipid droplets plays a role in pathogenesis. Dengue viral infections were shown to also increase the size and quantity of lipid droplets in infected cells (as much as a 3-fold increase) [25]. Cytoplasmic C protein is directed to lipid droplets by particular amino acids (spanning residues 46–66) on the hydrophobic $\alpha2$ helix [26]. In the study, substitution of two specific hydrophobic residues (L50S and L54S) in the $\alpha2$ helix resulted in a C protein that although was distributed in the cytoplasm, did not accumulate on the lipid droplet surfaces, highlighting that they are key to C protein-lipid droplet association [25].

Although viral translation, RNA replication, and encapsidation all occur in the cytoplasm of infected cells, the understanding of the coordination between the three processes is still murky. Artificial substitutions in L50S or L54S altered C protein targeting to the LDs, which resulted in normal RNA replication, but faulty production of an infectious particle, suggesting that viral assembly is affected [25]. The E protein, responsible for infectivity, was found to be reduced in these altered cells. When both L50S and L54S were mutated (a double mutant), also impaired infectious viral particle production, but also impaired amplification of viral RNA. This demonstrates that the when the C protein accumulates in the cytoplasm instead of being sequestered in the lipid droplets, it may prematurely interact with viral RNA, halting amplification of the viral RNA. A proposed biological function of the lipid droplet is that it acts as temporary storage site, sequestering the C protein, thereby controlling the amount of C protein available to interact with the viral RNA [37]. The understanding of the mechanism employed by the C protein to recruit viral RNA to form the nucleocapsid, as well as the location in which this occurs, is still uncertain. Two mechanisms have been proposed: (1) C protein is stored in the lipid droplet early in infection, and then deployed to the membrane of the endoplasmic reticulum where it will assemble with viral RNA to form nucleocapsids or (2) Nucleocapsids are produced at the site of C protein on the surface of the lipid droplets, and is then deployed to the membrane of the endoplasmic reticulum for further infectious viral assembly [25,33]. Nevertheless, localization of the C protein to the LDs has shown to be an integral part in infectious viral production, and thus can be a target for future eradication strategies for the Dengue virus. A study has shown that when a fatty acid synthase inhibitor was given, the amount of lipid droplets were reduced, leading to a substantial reduction in infectious viral particles produced [25].

## Capsid localization in the nucleus

In the nucleus, the C protein is primarily localized within nucleolus [38–40]. The role of the C protein in the nucleus is poorly understood, as its main function in nucleocapsid formation and virion assembly takes place in the cytoplasm. Recent studies have shown that within the nucleus of the infected cell, the C protein interacts with mostly nuclear host proteins. One such protein, the cellular protein nucleolin, is found in a larger extent in the nucleolus, with smaller amounts in the cytoplasm and plasma membrane. Nucleolin has a role in many cellular processes, including ribosome synthesis, protein transport, chromatin remodeling, control of translation, and RNA processing [41–44]. A study from 2013 demonstrated that the C protein-nucleolin interaction plays a significant role in formation of infectious viral particles (much like the lipid droplets). Researchers manipulated interactions between nucleolin and the C-protein, either through siRNA silencing of nucleolin, or with a nucleolin binding aptamer (AS1411), resulting in blunted interactions between the two. What was observed was a discernible reduction in viral titers, with no change in the intracellular Dengue viral RNA, C protein, or the E protein, indicating that viral RNA replication and synthesis of viral proteins were unaffected [45]. The C-protein-nucleolin interaction plays a larger role in virion morphogenesis, perhaps by the nucleolin protein acting as a chaperone for the C protein, aiding the formation of stable nucleocapsid particles. Other proteins that the C protein can interact within the nucleolus are DAXX and core histones. Binding of the C protein to DAXX was proposed to induce apoptosis via Fas, whereas binding to the core histones was proposed to disturb transcription by affecting the assembling of nucleosomes [46].

## Premembrane/membrane protein (prM/M) and envelope protein (E)

### *The precursor membrane-envelope protein complex*

The premembrane (prM) and the envelope (E) protein are acquired by the virion during budding off from the ER after nucleocapsid formation and along with lipids, forms the glycoprotein shell of the dengue viral particle. The E protein is the primary protein that is exposed to the immune system, to which antibodies are produced against during a Dengue viral infection. The E protein consists of an ectodomain (soluble E protein), a stem, and a transmembrane domain [47]. Both proteins have varying conformations depending on the maturity of the Dengue virus particle. In an immature Dengue virus particle, the prM and E together give a "spiky" appearance from the particle surface, made by 60 trimeric projections from 90 heterodimers [3]. During the transition through the Trans-Golgi Network, the spiky immature structure undergoes a conformational change into a mature, "smooth" appearance, a metamorphosis initiated by a low pH (5.8–6) [48–50] [illustrated in Fig. 6.5]. The E protein has a natural flexibility that allows the Dengue viral particle to undergo many conformational changes with changes in the environment.

Not only do the EDI-III regions differ in structure, they also differ greatly in procuring an immune response. The EDIII region generates robust neutralizing antibodies, which are usually serotype specific [51]. Serotype-specific antibodies are also generated against the EDI/EDII hinge zones and other complex epitopes on the E protein dimer [52]. These serotype-specific neutralizing antibodies are superior in preventing infection compared to cross-reactive nonneutralizing antibodies, which have been linked to causing antibody dependent enhancement (ADE) which leads to a more severe secondary infection with a different Dengue virus serotype [53]. The EDI and EDII regions also elicit antibodies, as does the prM protein, but these antibodies are weaker, nonneutralizing and frequently cross-reacting (especially anti prM antibodies). When the mature virus infects a new cell through clathrin-dependent receptor-mediated endocytosis, the E protein will undergo reconstruction to a trimer configuration in the endosome, and will protrude from the virus surface, allowing membrane fusion, allowing uncoating of the virion which will enable the release of viral RNA into the cytoplasm for subsequent translation. Therefore, the E protein can therefore undergo switching between three different oligomeric states: prM-E heterodimers in the immature Dengue virus particle, a dimer in the mature Dengue virus particle, and a trimer when the Dengue virus particle fuses with a host cell [47] [illustrated in Fig. 6.5].

## "SPIKY" immature virus

In the immature Dengue virus, the E and prM proteins together give a spiky appearance to the surface. This is made of 60 trimeric projections and 90 heterodimers. The E protein is capable of transforming from this immature configuration to a mature one due to its three domains that allow flexibility:

***EI:*** the central-most region of the E protein. It consists of an 8-stranded β-barrel. Is responsible for maintaining the structure of the E protein
***EII:*** EDII is dimer with 12 β strands, two α helices, and a fusion loop (FL). EII forms a hinge motion with EI.
***EIII:*** contains an immunoglobulin-like domain with 10 β-strands and functions in receptor binding to target cells. EIII also allows a hinge motion with its junction with EI. Also EIII generates the most robust antibodies.

The prM protein is unaffected by changes in pH, unlike the E protein. After the endoplasmic reticulum, within the trans-Golgi network the prM protein

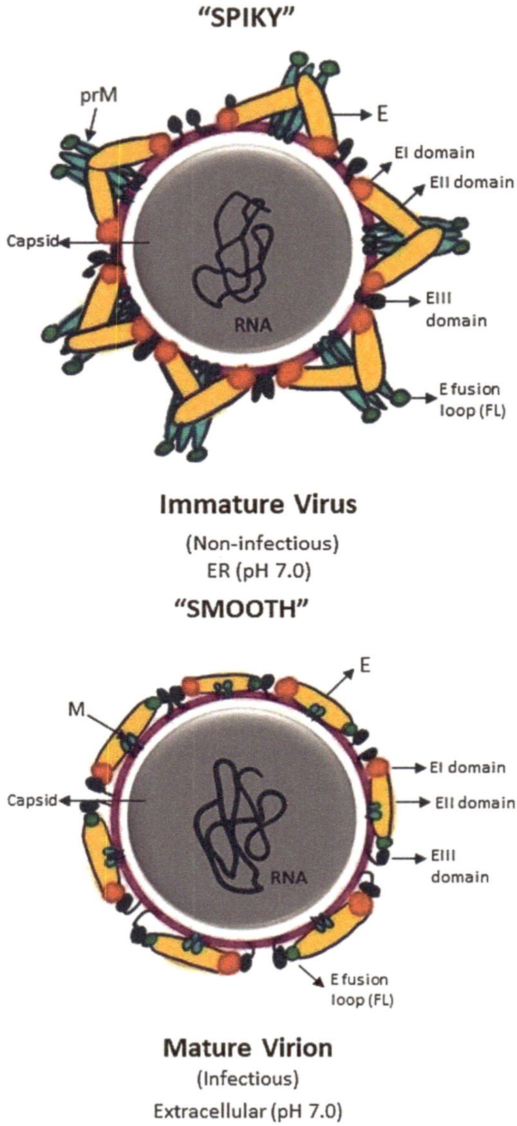


FIGURE 6.5 Conformational changes in the Dengue viral particle prM and E proteins. *prM*, premembrane; *M*, *membrane*; *E*, envelope; *EI*, Envelope domain one; *EII*, Envelope domain two; *EIII*, Envelope domain three; *FL*, Fusion loop; *ER*, Endoplasmic reticulum; *RNA*, Ribonucleic acid. *Illustration redrawn and adapted from Khetarpal N, Khanna I. Dengue Fever: Causes, Complications, and Vaccine Strategies. J Immunol Res. (2016).*

will be cleaved by furin which will yield a pr portion and an M portion. After cleavage, the mature virion with the E and the M proteins is excreted extracellularly.

## "SMOOTH" immature virus

During the transition of the Dengue virus through the trans-Golgi network, there is a conformational change from a spiky appearance to a smooth appearance, a process initiated by the low pH within the trans-Golgi network. The M and E complexes are now lying flat along the surface, as 90 homodimers. In the E homodimer, the monomers are arranged facing each other. If the pH rises at any point after this, the Dengue virion can once again switch to a spiky configuration, due to the flexibility of the E protein domains.

In contrast to the E protein, the tertiary structure of prM is unaffected by changes in pH. The prM protein consists of 166 amino acids, with seven antiparallel β-strands all stabilized with three disulfide bonds. While in the trans-Golgi network, the prM protein is cleaved by a host-encoded protease called furin that cleaves the prM protein into an N-terminal 91 amino acid "pr" peptide and an M protein. After cleavage, the "pr" peptide remains bound to the E protein forming a "cap," covering the hydrophobic fusion loops of all 3 E proteins, making the immature particle more hydrophilic. The pr peptide forming a cap on the E protein protects the fusion peptide on E from causing premature fusion of membranes, prior to mature virion release [50]. Following furin cleavage, the mature virion consisting of E and M are secreted extracellularly, and the pr peptide dissociates from the mature particle. Virion maturation is irreversible; after maturation the virion cannot revert back to its immature form, independently of pH. The average area of contact between the pr peptide and the E protein in an immature virus at a neutral pH was shown to be larger than the area of contact between the two in an immature virus at a lower pH (such as within the trans-Golgi network) [47]. This indicates that the pr protein has a greater affinity toward the E protein spike of an immature virus compared to the smooth surface of the mature virus. This reduction in affinity makes it easier for the pr protein to be released from the mature E structure at a neutral pH after cleavage by furin [47].

## Role of nonstructural proteins in viral replication and immune evasion

The NS proteins are synthesized after the structural proteins, following the E protein in the following order: NS1, NS2a, NS3, NS4a, NS4b, and NS5.

NS1 is the first NS protein to be synthesized. A 45kDA glycoprotein, it is highly conserved region, the structure differing only minutely among the flaviviruses. It contains two signals of Asn-X-Ser/Thr (where X is any amino acid) that is used for the insertion of an N-linked oligosaccharide [54]. It is synthesized initially as a hydrophilic monomeric glycoprotein on the rough endoplasmic reticulum, then shortly after switches to a homodimer that is more hydrophobic. Following dimer formation, the NS1 protein is then transferred to the lumen of the endoplasmic reticulum, followed by the Golgi apparatus, where its N-linked oligosaccharides are modified. This modification by the Golgi apparatus makes NS1 able to be transported to the cell membrane, remain intracellular, or secreted from the cell, although secretion of the NS1 is limited to mammalian cells and not seen in mosquitos [55]. It is involved in viral RNA replication, although the exact mechanism of how it does this is still unclear. Its role in viral defense is better understood, as it inhibits complement activation, thereby successfully evading cytolysis of infected cells [56].

NS2A/B are the next synthesized NS proteins. Unlike NS1 which is mostly secreted from the cell, both NS2A and NS2B remain predominately as transmembrane proteins. NS2A is 22 kDa primarily hydrophobic protein that is essential for proteolytic processing of the C terminus of the NS1 protein and is necessary for viral production [57]. Host signalases in the lumen of the endoplasmic reticulum cleave the transmembrane junction of NS2B-NS3, which causes the NS2A protein to split into two forms: NS2A and NS2Aα, both of which are necessary for proper viral replication and assembly. NS2B, after cleavage by host signalases, lies in the cytoplasm and is an essential cofactor for NS3, which is a viral serine protease, an RNA helicase, and an RTPase/NTPase [3]. When examining the Dengue viral two serotype, it was shown that residues of NS2B (40–80) are included in the NS3 protein, which were linked via a Gly-Ser linker to the N terminus of the full NS3 protein [3]. It was later discovered that the residues 67–80 specifically were crucial for the protease activity of the NS3 protein, due to NS2B encasing the protease domain of NS3, forming a fundamental portion of the protease active site of NS3 [58]. Hydrophobic parts of the NS2B protein also interact with NS3, such as the hydrophobic flanking regions that can anchor the NS2B-NS3 complex to the ER membranes [59]. This places the NS2B/NS3 protease complex in close proximity to the transmembrane portions of the polypeptide they are tasked with cleaving: The C-protein from the transmembrane anchor on the cytoplasmic side of the endoplasmic reticulum.

NS3 is a large, hydrophilic protein with a molecular weight of approximately 70,000 [54]. As stated previously, the NS3 complexes with the NS2B at the N-terminal region which functions as a serine protease which acts to cleave the polyprotein at the C protein-anchor site. The C-terminal site of NS3 resembles nucleoside-triphosphate binding protein sequences, therefore, playing a role in Dengue viral RNA replication. The NS3 protein is one of the most important proteins involved in Dengue pathogenesis, with three main functions: (1) Protease activity which cleaves the viral polyprotein, (2) RNA helicase activity and RTPase/NTPase-viral RNA replication, and (3) induces apoptosis in virally infected Dengue cells [31]. The helicase portion of the

NS3 protein (residues 180–618), has three subdomains very similar in structure to others flaviviruses. The subdomains I and II (residues 181–326, residues 327–481, respectively), has six parallel β-pleated sheets, with four α-helices on the ends of the sheets, a configuration very similar to the hepatitis C virus (HCV) helix structure [60,61]. The last subdomain, subdomain III (residues 482–618) is more unique in dengue virus, composed of four parallel α-helices, enclosed by three α-helices and two antiparallel β strands. The helicase activity of the NS3 protein is thought to be influenced by the protease activity of NS3, as helicase activity was found to be significantly enhanced (about 30-fold) in the full-length NS3 protein compared to the solitary helicase domain of the NS3 protein [62]. This suggests that the protease domain is linked to the helicase domain via a linker molecule that allows the protease domain to influence and augment helicase activity. The NS3 helicase domain is also suggested to interact with the NS5 protein, which is the RNA polymerase responsible for viral RNA replication [63]. The NS3 protein also plays a role in inducing apoptosis, and recently has shown to also play a significant role in viral assembly that takes place after viral RNA replication [64,65]. Overall, the NS3 protein is a multifaceted, involved in multiple processes within the infected cell, thereby playing a vital role in viral morphogenesis.

NS4A and NS4B, 16 kDa and 27 kDA, respectively, are membrane proteins, and are responsible for inducing membrane alterations necessary for the replicative process of the Dengue virus. The NS4A protein predominately induces membrane alterations and enhances autophagy (destruction of cellular components in vacuoles), both of which augment viral RNA replication [66,67]. NS4B assists in viral replication by its interaction with the NS3 protein, a protein vital for viral replication via helicase activity [68]. NS4B also plays a role in immune evasion, by blocking IFN α/β signal transduction via the JAK/STAT pathway, which allows the virus to circumvent the host's innate immune response to viral infections [69].

The last NS protein, NS5, is the largest of the NS proteins, with a molecular weight of 104,000 and composed of 900 amino acid residues. It is also the most conserved of the NS proteins among the Dengue serotypes, with a 67% sequence homology among the four serotypes [3]. Residues 1–296 at the N-terminus make up the methyltransferase domain of the protein, while the C-terminus consists of an RNA-dependent RNA-polymerase (RdRp), which is composed of residues 320–900. The RdRP domain is similar in structure among the flaviviruses, with a canonical conformation with palm, fingers, and thumb subdomains [70]. It contains a GDD motif, common among flaviviruses, which is responsible for catalytic activity responsible for adding nucleotides using metal ions. "Finger tips" connect the finger and thumb domains to form an active site, a feature that differentiates the RdRps from DNA-dependent RNA polymerases [3]. Between residues 320–405, RdRps have a nuclear localization signal (NLS), as NS5 protein is predominately localized within the nucleus in Dengue viral infections, a feature not shared by flaviviruses [71,72]. Residues 320–368 specifically are suggested to play a role in the interaction of the NS5 protein with the NS3 protein (the helicase domain) [63]. The localization of the NS5 protein within the nucleus is an oddity, as its main function (RdRp) is required for viral replication within the cytoplasm. The reasoning behind this is unclear, but suggests that in addition to its enzymatic activities, the NS5 protein perhaps interacts with host proteins to modulate viral morphogenesis [73]. The methyltransferase domain at the N-terminal end of the NS5 protein is composed of a core consisting of S-adenosyl-methionine-dependent MTase (SAM-dependent MTAse), that is folded into an α/β/β structure between the N and C terminal regions of NS5 [3]. The methyltransferase activity of this region is responsible for the synthesis of the methyl-guanosine cap at the 5′ end of the newly synthesized viral RNA during viral replication. The SAM-binding site of the NS5 region has shown to have a more open configuration, when compared to the West Nile Virus [74,75]. This increase openness suggests that the byproduct of methylation, the S-adenosyl-homocysteine (SAH) is more readily released from the cell.

## Cis-acting RNA structures that influence viral replication

The highly structured 5′ and 3′ untranslated regions (UTRs) as well as 5′ and 3′ coding regions contain RNA elements that are responsible for the initiation of translation and can act to enhance, suppress, or mediate replication of the Dengue viral genome.

### 5′ UTR and 5′ coding regions

The viral 5′ UTR of the Dengue virus genome, around 100 nucleotides long, is vital for proper viral replication, with a cap-like structure on the 5′ end that is highly conserved among the flaviviruses [9]. The 5′ UTR region contains three essential regions for viral replication: (1) a promoter region called the stem-loop A (SLA) that binds and activates viral RNA polymerase [76], (2) stem-loop B (SLB), which has a sequence known as 5′ upstream of the AUG (start codon) region (5′ UAR) that is complementary to the 3′ end of the viral genome, responsible for the long RNA–RNA interactions causing cyclization, and lastly (3) A region rich in U nucleotides between SLA and SLB that acts as an enhancer of replication [77]. Additionally, there are regions in the 5′ coding region, especially in the coding sequence of the viral capsid protein, that play a significant role in mediating long range RNA–RNA interactions. These include: (1) a 5′ cyclization sequence (5′CS) and (2) a 5′ downstream of AUG region (5′DAR) both of which are located in the capsid coding region, and (3) a hairpin located between the 5′CS and the 5′DAR called cHP [76]. Cyclization of the flavivirus genome is proposed to be essential for viral RNA replication, possible by complementary sequences present at the 5′ and 3′ ends of the genome (depicted in Fig. 6.6].

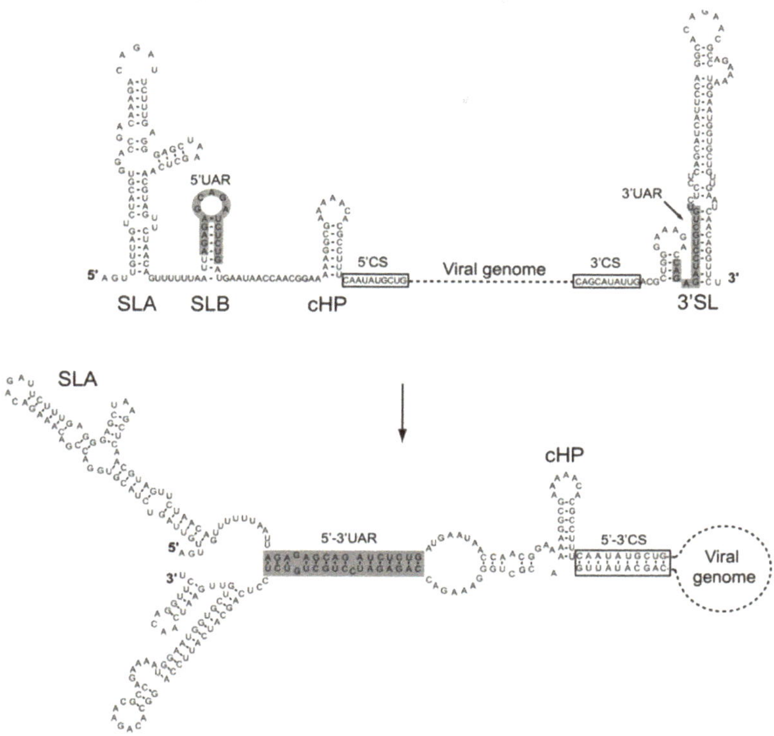


**FIGURE 6.6** Sequences located at the 5′ and 3′ ends of the Dengue Virus genome. *SLA*, stem loop A; *SLB*, stem loop B; *UAR*, upstream of the AUG region. *Illustration from Villordo S M et al. Genome Cyclization as Strategy for Flavivirus RNA Replication. Virus Res (2017).*

## 3′ UTR region

The 3′ UTR region is a 450-nucleotide-long region that lacks a poly-A tail. Instead, it ends in a highly conserved 3′ stem loop (3′SL), a structure conserved not only among different Dengue virus serotypes, but also among the flaviviruses [78]. This 3′SL region is essential for viral replication. Upstream of the 3′SL is a region known as the conserved sequence (CS1), which contains a cyclization sequence (3′CS). Both these regions of the 3′UTR have complementary sequences to areas in the 5′ UTR and 5′ coding sequences, which allows for cyclization of the viral genome, bringing the 3′ end closer to the RNA polymerase at the promoter site of the 5′ end [79]. Cyclization of the flavivirus genome is proposed to be essential for viral RNA replication. There is also a variable region (VR) downstream of the ORF. It has been suggested that this VR sequence may have a similar function to the poly-A tail, due to studies that have shown similar efficiencies in translation between mRNAs with a VR sequence and mRNAs with poly-A tails [78]. Deletions in the VR region can lead to a reduction in viral replication in vivo [80]. In one study studying the Dengue virus 2 serotype, it was shown that a 10-nucleotide deletion was linked to dengue clinical presentation, resulting in a milder clinical picture [81]. Located just downstream to the VR region and just upstream of the 3′SL, are the A2 and A3 regions of the 3′UTR. These regions contain highly conserved sequences CS2 within the A3 region and repeated CS2 (RCS2) within the A2. While deletions in either A2 or A3 had no significant effect on viral translation, concomitant deletions in both A2 and A3 decreased viral translation by 30%–40%, suggesting A2-A3 may play a role in stabilizing RNA. Individual deletions in either A2 or A3 while showing only trivial reductions in viral translation, showed large reductions in viral RNA synthesis, an effect that was augmented by a deletion in both [82].

### *Cyclization of the viral genome*

Cyclization of the viral genome is mediated by complementary sequences in the 5′ end of the viral genome (both 5′ UTR and 5′ coding regions) and the 3′ UTR. There are two main complementary sequences that hybridize required for cyclization: (1) between the 3′SL and the 5′ UAR located in the 5′ UTR and (2) between the 3′CS to a region in the coding sequence for capsid protein in the 5’ [21]. Mutational studies in these regions demonstrated that these interactions are essential for viral RNA synthesis, as there was a notable reduction in viral RNA synthesis (as much as 200-300-fold) when these areas were mutated or deleted [77]. Single mutations that disturbed the 5′-3′ UAR hybridization, were shown to drastically reduce synthesis of viral RNA, in some cases to subdetectable levels. Mutations that restored the base pairings in these regions, were shown to “recover” the base pairings, and thus RNA synthesis occurred normally [77]. Single nucleotide mutations in the 5′ UAR region that did not affect the complementary sequences, were also shown to better tolerated, highlighting the importance of these complementary sequences for effective viral replication. The 3′SL region was suggested to alter in structure upon binding to the 5′ UAR region. One such proposed mechanism suggested that terminal nucleotides in the 3′ end were structurally changed (“melted”), and that triggered the initiation of RNA polymerase activity [83–85]. Further studies are warranted to determine if the hybridization of the 5′UAR region and the 3′SL region is essential in only bringing the 3′ end in close proximity to the polymerase at the 5′ end, or if it is necessary for triggering a conformational change in the 3′SL that initiates viral RNA polymerase activity. Alterations in the SLB stem region of the 5′ UTR not including the 5′ UAR region were shown to have no effect of viral RNA synthesis [77]. The stem of the SLB region was thought to assist in maintaining UAR complementarity, but the unaffected viral translation after mutations, suggest that the SLB stem can play a role in other parts of viral morphogenesis, such as viral assembly.

## *Host specification by RNA duplication regions in the 3′UTR*

RNA structures that arise from duplications within the 3′ UTR known as dumbbells (DB1 and DB2) have shown to play a role in host switching between mosquito and human cells. These dumbbells contain sequences that overlap with the essential RNA sequences in the 3′ UTR responsible for cyclization, and DB1 and DB2 are thought to have differing functions regulating genome cyclization [86]. The dumbbell regions were initially thought to enhance viral replication in both mosquitos and humans, but recent research has discovered that these regions greatly enhance or diminish viral replication in mosquito cells only, while having a negligible effect on viral replication in humans [87]. This demonstrates that these RNA structures in the 3′ UTR are under different pressures depending on the host infected by the Dengue virus. Not only are there two dumbbell elements (DB1 and DB2), there is also nearly identical stem-loop structures (SLI and SLII), elements not to be confused with the 3′ stem loop (3′SL) [86]. These pairs of duplicated RNA elements (SLI-SLII and DB1-DB2) interact with an area called the pseudoknot (PK), which is able to hinder degradation of the viral genome. In the mosquito, mutations in the DB2 sequence, enhanced viral replication. Sequences in the D1 region hybridize with a sequence in the 5′ capsid coding sequence, enhancing genomic cyclization. The DBs have contrasting effects: DB2 complexes with PK forming a DB2-PK complex that reduces long-range RNA—RNA interactions responsible for cyclization, and therefore reduce viral replication, while DB1 promoted RNA—RNA interactions and cyclization [88]. This contrasting nature shows that mutations in DB1 or DB2 yield different effects on viral RNA production. Complete deletions of DB2 resulted in an increase in viral replication in mosquito cells [86].

Understanding these DB structures and their effect on viral replication in infected mosquito cells, suggests that they can be targets for future therapies to reduce Dengue viral transmission to human hosts. Another structure, a small hairpin-like structure (sHP) in the 3′ UTR, lying just upstream form the terminal 3′SL, is an essential region for Dengue viral replication in both hosts. A sequence in the loop of the sHP was found in recent studies to play a vital role in replication, but only in mosquito cells, while being expendable in human hosts [86].

## *Subgenomic flavivirus RNAs (sfRNAs)*

Subgenomic flavivirus RNAs (sfRNAs) are derived from the 3′ UTR and are capable of blunting the host antiviral response (innate immunity) in both mammalian cells as well as mosquito cells, increasing the epidemiological fitness of the Dengue virus [89]. During Dengue viral replication, a cellular 5′-3′ exoribonuclease (XRN1) "trims" the viral genomic RNA, digesting it until the conserved stem loop-II (SL-II) structure, which stalls exoribonuclease activity. SL-II is a structure in the 3′ UTR that interacts with the pseudoknot

(PK) also in the 3′UTR. This results in the production of a subgenomic flavivirus RNA (sfRNA) ($\approx$0.5 kb) [90]. XRN1 also stalls at SL-IV and DB1 RNA structures in the 3′UTR, forming a smaller sfRNA2 and sfRNA3 [90]. One of the initial responses to a viral infection involves the production of interferons (IFNs), especially IFN-$\alpha$ and IFN-$\beta$, which are responsible for the cytolysis of Dengue virus infected cells. sfRNAs interact with host proteins, functioning to suppress translation of the IFN genes [91].

The innate immune response against flaviviruses is initially triggered by retinoic acid-inducible gene-I-like receptors (RIG-I, MDA5) sensing viral RNA [90]. The binding of the RIG-I receptors to the viral RNA will prompt transcription of the interferon response factor (IRF) which will ultimately lead to the induction of interferon (IFN). A recent study exploring the functional intricacies of sfRNAs demonstrated a binding of sfRNA from the Dengue virus 2 serotype to the ubiquitin ligase tripartite motif protein 25 (TRIM25) [90]. TRIM25 functions normally to polyubiquitynate the RIG-I receptor, which is essential for RIG-I signaling a vigorous response when encountering viral RNA. The strong binding of sfRNA to TRIM25 prevents the function of TRIM25, preventing ubiquitination of RIG-I [90]. sfRNA also represses innate immune responses indirectly. It was shown to interact with stress granules (in the ER), especially components G3BP1/2 and Caprin, which subsequently decreased IFN-stimulated gene (ISG) translation [90].

## ER stress and the unfolded protein response (UPR): implications in viral pathogenesis

Dengue virus utilizes the host machinery for the synthesis of viral proteins that contribute to its pathogenicity. Dengue virus infections cause significant stress in the endoplasmic reticulum which triggers a cellular reaction in infected cells called the unfolded protein response, which is activated to relieve stress in the endoplasmic reticulum [92]. In Dengue virus, this response allows for cellular survival and expedites removal of the virus. Dengue virus can adapt to alter this host response to its own advantage, interacting with complex cellular reactions such as autophagy, apoptosis, innate immune responses, and proinflammatory reactions [93,94]. Its modifications of these host cellular processes allows Dengue virus to escape immune surveillance and continue viral replication within cells. The overproduction of unfolded or misfolded proteins are a normal consequence during viral/bacterial infections and overwhelm the ability of the endoplasmic reticulum to handle them [94,95]. This resulting stress in the endoplasmic reticulum caused by the unfolded/misfolded proteins triggers the onset of the unfolded protein response for cell survival. There are three separate pathways activated by the unfolded protein response (depicted in Fig. 6.4) that are induced by Dengue infection. The pathways ultimately result in: (1) suppression of mRNA translation (halts protein synthesis), (2) an increase in endoplasmic reticulum protein folding capacities, (3) endoplasmic reticulum-associated degradation (ERAD) of misfolded proteins that have accumulated in the endoplasmic reticulum, (4) increased synthesis of chaperones in the endoplasmic reticulum, (5) inducing a proinflammatory state within the cell, and (6) inducing apoptosis of the cell [92].

## Pathways utilized by the unfolded protein response

### *Inositol requiring kinase 1 (IRE1) pathway*

When IRE1 is intrinsically autophoshorylated, it will result in mRNA and miRNA degradation in a manner called regulated IRE1-dependent decay (RIDD). mRNAs produced will also then induce ER-associated degradation (ERAD) genes to degrade misfolded proteins. Dengue virus activates this pathway, and further activates cellular caspases (3 and 9) to trigger cellular apoptosis [92] [Summarized in Fig. 6.7].

### *Protein kinase R-like ER kinase (PERK) pathway*

Phosphorylated PERK will inhibit the eukaryotic translation initiator factor 2$\alpha$ (eIF2$\alpha$), resulting in a blockage of protein translation. eIF2$\alpha$ also activates a transcription factor ATF-4, which will further activate a C/EBP protein (CHOP) and a protein called the growth arrest and DNA damage-inducible protein (GADD34), both of which will trigger apoptosis [92]. [Summarized in Fig. 6.7].

### *Activating transcription factor-6 (ATF6) pathway*

ATF6 when activated, will result in the uncovering of a Golgi localization signal. Cleavage of the ATF6 in the Golgi apparatus, will yield an activated transcription factor, ATF6p50, which upon translocating to the nucleus, will result in the increased transcription of genes related to ERAD [92] [Summarized in Fig. 6.7].

## Manipulation of the unfolded protein response by the Dengue virus

### *Viral replication enhancement*

The modulation of the unfolded protein response by the Dengue virus is complex, with the PERK arm of the pathway activated early in disease, followed by the IRE1-XBP1, and lastly the ATF6 late in the infection [96]. This allows the Dengue virus to overcome the translation attenuation created by the unfolded protein response, allowing it to continuously produce viral proteins necessary for viral replication (such as NS5 which functions as the RNA polymerase). Dengue virus is also able to inhibit apoptosis by blocking the downstream signaling of IRE1-XBP1 that normally activates caspases, allowing for uninterrupted dengue viral replication [96].

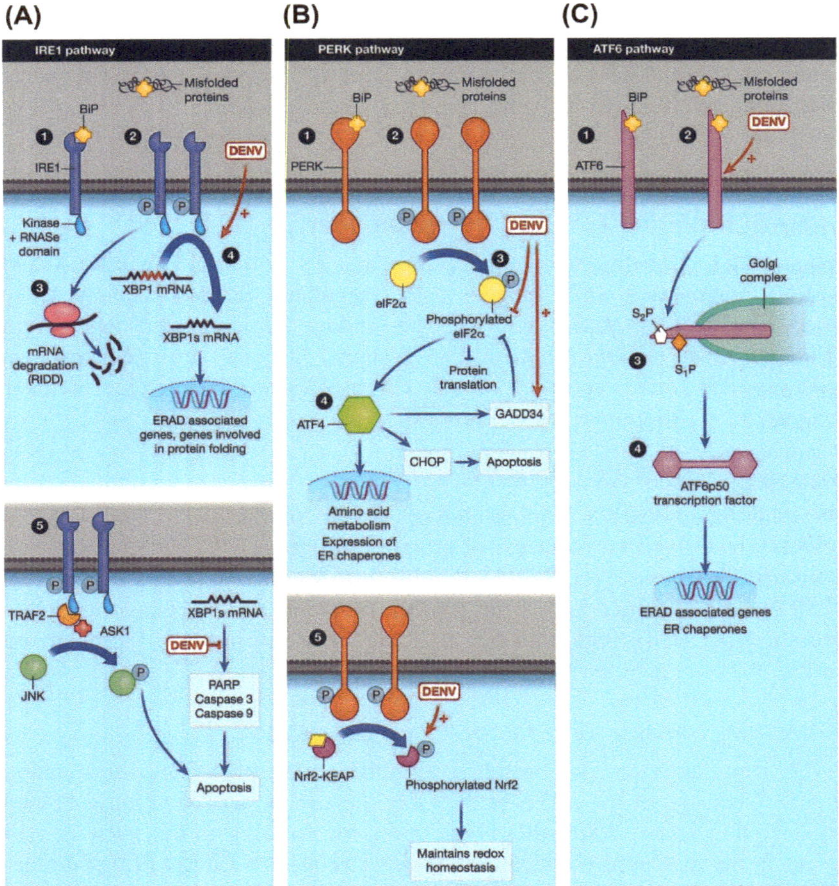


FIGURE 6.7 **Unfolded protein response and ER stress:** Modulation by the Dengue Virus. (A) IRE1 pathway: Dengue virus will activate IRE1 phosphorylation which will result in mRNA degradation. Dengue virus also activates caspases (caspase three and 9), (B) PERK pathway: Dengue virus phosphorylates eIF2α and activates GADD34 to trigger apoptosis, (C) ATF6 pathway: Dengue virus causes activation of ATF6p50, a transcription factor that increases transcription of genes related to ER stress pathways. *ER: endoplasmic reticulum. Illustration from Perera N, Miller JL et al. The role of the unfolded protein response in dengue virus pathogenesis. Cell Microbio (2017).*

## Autophagy modulation

In viral infections, autophagy plays a significant role in removing viral components and allows for viral antigen presentation to augment an immune response. Dengue virus manipulates this process by inducing autophagy in the cell, allowing sequestration of Dengue virus within the double membrane vesicles that are normally produced in the autophagic process. Isolation of Dengue virus in these vesicles allows for not only viral replication going unhindered, but also evades the antibody-dependent neutralization against viral antigens [97,98]. Another advantage to activating autophagy is that β-oxidation of lipids occur, which provides ATP necessary for viral replication [99].

### Inflammation induced by Dengue virus

Dengue viral infections normally trigger activation of the unfolded protein response, which will then activate innate proinflammatory immune mediators such as PKR, IRF3, IL-1β, and NF-κB, which function to eradicate Dengue virus [93]. It is a double-edged sword however, as too much of a robust inflammatory response can cause harm to the host, as seen with the phenomenon of increased vascular permeability in severe Dengue infections. The type I IFN response is a vital antiviral mechanism, to which dengue virus has adapted strategies to evade. Dengue virus not only inhibits the induction of IFN, but also its downstream JAK-STAT signaling to ensure its survival within infected cells [100]. While evading the innate immune response (IFN), Dengue virus additionally increases the production of proinflammatory cytokines, such as TNF-α. This results in essentially a "cytokine storm" that can cause the grave clinical presentations seen in severe Dengue infections [92].

## Conclusion

Structural analysis of the Dengue virus has certainly come a long way in the recent years and has allowed for a much-needed perceptivity regarding the complex pathogenesis of Dengue viral infections. While great strides have been made in understanding the Dengue virus genomic structure, the presence and functions of viral structural and NS proteins, and the Dengue virus life cycle, Dengue virus is very much still on the rise today, posing a significant global health risk. Dengue virus is an astute one, capable of manipulating the host's response in its favor, making mechanisms to eradicate or control it quite challenging. Thankfully, recent advancements in understanding the intricacies of the Dengue viral structure and genome has allowed a much more nuanced picture on how and why Dengue virus does what it does in the host cell. Knowing the complexities of the Dengue viral structure and genetics allows for new and innovative ways to combat disease, a feat we are optimistic for in the near future.

## References

[1] Khetarpal N, Khanna I. Dengue fever: causes, complications, and vaccine strategies. J Immunol Res 2016;2016:6803098. https://doi.org/10.1155/2016/6803098 [Epub 2016 Jul 20].

[2] Zhang X, Ge P, Yu X, Brannan JM, Bi G, et al. Cryo-EM structure of the mature dengue virus at 3.5-Å resolution. Nat Struct Mol Biol 2013;20:105–10 [PubMed: 23241927].

[3] Perera R, Kuhn RJ. Structural proteomics of dengue virus. Curr Opin Microbiol 2008;11(Issue 4):369–77. https://doi.org/10.1016/j.mib.2008.06.004. ISSN 1369-5274.

[4] Diamond MS, Pierson TC. Molecular insight into dengue virus pathogenesis and its implications for disease control. Cell 2015;162(3):488–92. https://doi.org/10.1016/j.cell.2015.07.005.

[5] Byk LA, Iglesias NG, De Maio FA, Gebhard LG, et al. Dengue virus genome uncoating requires ubiquitination. mBio June 28, 2016;7(3):e00804–16. https://doi.org/10.1128/mBio.00804-16.

[6] Byk LA, Gamarnik AV. Properties and functions of the dengue virus capsid protein. Annu Rev Virol September 29, 2016;3(1):263–81. Epub 2016 Aug 3.

[7] Ma L, Jones CT, Groesch TD, Kuhn RJ, Post CB. Solution structure of dengue virus capsid protein reveals another fold. Proc Natl Acad Sci USA 2004;101:3414–9 (PubMed: 14993605).

[8] Nemesio H, Palomares-Jerez MF, Villalain J. Hydrophobic segment of dengue virus C protein. Interaction with model membranes. Mol Membr Biol 2013;30:273–87 (PubMed: 23745515).

[9] Markoff L. 5'-3'-noncoding regions in flavivirus RNA. Adv Virus Res 2003;59:177–228.

[10] Khromykh AA, Westaway EG. RNA binding properties of core protein of the flavivirus Kunjin. Arch Virol 1996;141:685–99 (PubMed: 8645104).

[11] Ivanyi-Nagy R, Lavergne JP, Gabus C, Ficheux D, Darlix JL. RNA chaperoning and intrinsic disorder in the core proteins of *Flaviviridae*. Nucleic Acids Res 2008;36:712–25 (PubMed: 18033802).

[12] Lobigs M, Lee E, Ng ML, Pavy M, Lobigs P. A flavivirus signal peptide balances the catalytic activity of two proteases and thereby facilitates virus morphogenesis. Virology 2010;401:80–9 (PubMed: 20207389).

[13] Stocks CE, Lobigs M. Signal peptidase cleavage at the flavivirus C-prM junction: dependence on the viral NS2B-3 protease for efficient processing requires determinants in C, the signal peptide, and prM. J Virol 1998;72:2141–9 (PubMed: 9499070).

[14] Lobigs M, Lee E. Inefficient signalase cleavage promotes efficient nucleocapsid incorporation into budding flavivivirus membranes. J Virol 2004;78:178–86 (PubMed: 14671099).

[15] Welsch S, Miller S, Romero-Brey I, Merz A, Bleck CK, et al. Composition and three-dimensional architecture of the dengue virus replication and assembly sites. Cell Hostw Microbe 2009;5:365–75 (PubMed: 19380115).

[16] Gillespie LK, Hoenen A, Morgan G, Mackenzie JM. The endoplasmic reticulum provides the membrane platform for biogenesis of the flavivirus replication complex. J Virol 2010;84:10438–47 (PubMed: 20686019).

[17] Junjhon J, Pennington JG, Edwards TJ, Perera R, Lanman J, Kuhn RJ. Ultrastructural characterization and three-dimensional architecture of replication sites in dengue virus-infected mosquito cells. J Virol 2014;88:4687–97 (PubMed: 24522909).

[18] Khromykh AA, Varnavski AN, Sedlak PL, Westaway EG. Coupling between replication and packaging of flavivirus RNA: evidence derived from the use of DNA-based full length cDNA clones of Kunjin virus. J Virol 2001;75:4633–40 (PubMed: 11312333).

[19] Schalich J, Allison SL, Stiasny K, Mandl CW, Kunz C, Heinz FX. Recombinant subviral particles from tick-borne encephalitis virus are fusogenic and provide a model system for

studying flavivirus envelope glycoprotein functions. J Virol 1996;70:4549–57 (PubMed: 8676481).

[20] Scaturro P, Cortese M, Chatel-Chaix L, Fischl W, Bartenschlager R. Dengue virus non-structural protein 1 modulates infectious particle production via interaction with the structural proteins. PLoS Pathog 2015;11:e1005277 (PubMed:26562291).

[21] deBorba L, Villordo SM, Iglesias NG, Filomatori CV, Gebhard LG, Gamarnik AV. Overlapping local and long-range RNA-RNA interactions modulate dengue virus genome cyclization and replication. J Virol 2015;89:3430–7 (PubMed: 25589642).

[22] Clyde K, Barrera J, Harris E. The capsid-coding region hairpin element (cHP) is a critical determinant of dengue virus and West Nile virus RNA synthesis. Virology 2008;379:314–23 (PubMed: 18676000).

[23] Liu ZY, Li XF, Jiang T, Deng YQ, Zhao H, et al. Novel *cis*-acting element within the capsid-coding region enhances flavivirus viral-RNA replication by regulating genome cyclization. J Virol 2013;87:6804–18 (PubMed: 23576500).

[24] Samsa MM, Mondotte JA, Caramelo JJ, Gamarnik AV. Uncoupling *cis*-acting RNA elements from coding sequences revealed a requirement of the N-terminal region of dengue virus capsid protein in virus particle formation. J Virol 2012;86:1046–58 (PubMed: 22072762).

[25] Samsa MM, Mondotte JA, Iglesias NG, Assuncao-Miranda I, Barbosa-Lima G, et al. Dengue virus capsid protein usurps lipid droplets for viral particle formation. PLoS Pathog 2009;5:e1000632 (PubMed: 19851456).

[26] Markoff L, Falgout B, Chang A. A conserved internal hydrophobic domain mediates the stable membrane integration of the dengue virus capsid protein. Virology 1997;233:105–17 (PubMed:9201220).

[27] Patkar CG, Jones CT, Chang YH, Warrier R, Kuhn RJ. Functional requirements of the yellow fever virus capsid protein. J Virol 2007;81:6471–81 (PubMed: 17526891).

[28] Kofler RM, Heinz FX, Mandl CW. Capsid protein C of tick-borne encephalitis virus tolerates large internal deletions and is a favorable target for attenuation of virulence. J Virol 2002;76:3534–43 (PubMed: 11884577).

[29] Zhu W, Qin C, Chen S, Jiang T, Yu M, et al. Attenuated dengue 2 viruses with deletions in capsid protein derived from an infectious full-length cDNA clone. Virus Res 2007;126:226–32 (PubMed: 17412442).

[30] Schlick P, Taucher C, Schittl B, Tran JL, Kofler RM, et al. Helices α2 and α3 of West Nile virus capsid protein are dispensable for assembly of infectious virions. J Virol 2009;83:5581–91 (PubMed: 17412442).

[31] Mackenzie JM, Jones MK, Young PR. Immunolocalization of the dengue virus nonstructural glycoprotein NS1 suggests a role in viral RNA replication. Virology 1996;220:232–40 (PubMed: 8659120).

[32] Miorin L, Romero-Brey I, Maiuri P, Hoppe S, Krijnse-Locker J, et al. Three-dimensional architecture of tick-borne encephalitis virus replication sites and trafficking of the replicated RNA. J Virol 2013;87:6469–81 (PubMed: 23552408).

[33] Thiele C, Spandl J. Cell biology of lipid droplets. Curr Opin Cell Biol 2008;20:378–85.

[34] Moradpour D, Englert C, Wakita T, Wands JR. Characterization of cell lines allowing tightly regulated expression of hepatitis C virus core protein. Virology 1996;222:51–63.

[35] Barba G, Harper F, Harada T, Kohara M, Goulinet S, et al. Hepatitis C virus core protein shows a cytoplasmic localization and associates to cellular lipid storage droplets. Proc Natl Acad Sci USA 1997;94:1200–5.

[36] Hope RG, McLauchlan J. Sequence motifs required for lipid droplet association and protein stability are unique to the hepatitis C virus core protein. J Gen Virol 2000;81:1913–25.
[37] Cermelli S, Guo Y, Gross SP, Welte MA. The lipid-droplet proteome reveals that droplets are a protein-storage depot. Curr Biol 2006;16:1783–95.
[38] Bulich R, Aaskov JG. Nuclear localization of dengue 2 virus core protein detected with monoclonal antibodies. J Gen Virol 1992;73:2999–3003.
[39] Sangiambut S, Keelapang P, Aaskov J, Puttikhunt C, Kasinrerk W, et al. Multiple regions in dengue virus capsid protein contribute to nuclear localization during virus infection. J Gen Virol 2008;89:1254–64.
[40] Wang S-H, Syu W-J, Huang K-J, Lei H-Y, et al. Intracellular localization and determination of a nuclear localization signal of the core protein of dengue virus. J Gen Virol 2002;83:3093–102.
[41] Mongelard F, Bouvet P. Nucleolin: a multifaceted protein. Trends Cell Biol 2007;17:80–6.
[42] Tajrishi MM, Tuteja R, Tuteja N. Nucleolin: the most abundant multifunctional phosphoprotein of nucelolus. Commun Integr Biol 2011;4:267–75.
[43] Lo SJ, Lee C-C, Lai H-J. The nucleolus: reviewing oldies to have new understandings. Cell Res 2006;16:530–8.
[44] Abdelmohsen K, Gorospe M. RNA-binding protein nucleolin in disease. RNA Biol 2012;9:799–808.
[45] Balinsky C, Schmeisser H, Ganesan S, Singh K, et al. Nucleolin interacts with the dengue virus capsid protein and plays a role in formation of infectious virus particles. J Virol 2013:00704–13.
[46] Mairiang D, Zhang H, Sodja A, Murali T, Suryiyphol P, et al. Identification of new protein interactions between dengue fever virus and its hosts, human and mosquito. PLoS One 2013;8(1):e53535. https://doi.org/10.1371/journal.pone.0053535.
[47] Li L, Lok SM, Yu IM, Zhang Y, Kuhn RJ, Chen J, Rossman MG. The flavivirus precursor membrane-envelope protein complex: structure and maturation. Science 2008;319(5871):1830–4. https://doi.org/10.1126/science.1153263.
[48] Modis Y, Ogata S, Clements D, Harrison SC. Structure of the dengue virus envelope protein after fusion. Nature 2004;427:313–9 (PubMed: 14737159).
[49] Zhang Y, Zhang W, Ogata S, Clements D, et al. Conformational changes of the flavivirus E glycoprotein. Structure 2004;12:1607–18 (PubMed: 15341726).
[50] Yu I-M, Zhang W, Holdaway HA, Li L, et al. Structure of immature dengue virus at low pH primes proteolytic maturation. Science 2008;319(5871):1834–7.
[51] R. de Alwis, M. Beltramello, W.B Messer, et al. In Depth analysis of the antibody response of individuals exposed to primary dengue virus infection. PLoS Neglected Trop Dis, vol. 5, no. 6, Article ID e1188, 2011
[52] B Messer W, de Alwis R, Yount BL, et al. Dengue virus envelope protein domain I/II hinge determines long-lived serotype-specific dengue immunity. Proc Natl Acad Sci USA 2014;III(5):1939–44.
[53] Dejnirattisai W, Jumnainsong A, Onsirisakul N, et al. Cross-reacting antibodies enhance dengue virus infection in humans. Science 2010;328(5979):745–8.
[54] Henchal E, Putnak R. The dengue viruses. Clin Microbiol Rev 1990:276–396.
[55] Mason PW. Maturation of Japanese encephalitis virus glycoproteins produced by infected mammalian and mosquito cells. Virology 1989;169:354–64.

[56] Rice CM, Strauss EG, Strauss JH. Structure of the flavivirus genome. In: Schlesinger S, Schlesinger M, editors. Togaviruses and flaviviruses. New York: Plenum Publishing Corp.; 1986. p. 279–327.

[57] Falgout B, Chanock R, Lai CJ. Proper processing of dengue virus nonstructural glycoprotein NS1 requires the N-terminal hydrophobic signal sequence and the downstream nonstructural protein, NS2a. J Virol 1989;63:1852–60.

[58] Erbel P, Schiering N, D'Arcy A, Renatus M, Kroemer M, Lim SP, Yin Z, et al. Structural basis for the activation of flaviviral NS3 proteases from dengue and West Nile virus. Nat Struct Mol Biol 2006;13:372–3 (PubMed: 16532006).

[59] Clum S, Ebner KE, Padmanbhan R. Cotranslational membrane insertion of the serine proteinase precursor NS2B-NS3 (Pro) of dengue virus type 2 is required for efficient in vitro processing and is mediated through the hydrophobic regions of NS2B. J Biol Chem 1997;272:30715–23 (PubMed: 9388208).

[60] Yao N, Hesson T, Cable M, Hong Z, Kwong AD, Le HV, et al. Structure of the hepatitis C virus RNA helicase domain. Nat Struct Biol 1997;4:463–7 (PubMed: 918754).

[61] Kim JL, Morgenstern KA, Griffith JP, Dwyer MD, Thomson JA, et al. Hepatitis C virus NS3 RNA helicase domain with a bound oligonucleotide: the crystal structure provides insights into the mode of unwinding. Structure 1998;6:89–100 (PubMed: 9493270).

[62] Wu J, Bera K, Kuhn RJ, Smith JL. Structure of the flavivirus helicase: implications for catalytic activity, protein interactions, and proteolytic processing. J Virol 2005;79:10268–77 (PubMed: 16051820).

[63] Kapoor M, Zhang L, Ramachandra M, Kusukawa J, et al. Association between NS3 and NS5 proteins of dengue virus type 2 in the putative RNA replicase is linked to differential phosphorylation of NS5. J Biol Chem 1995;270:19100–6 (PubMed: 7642575).

[64] Patkar CG, Kuhn RJ. Yellow fever virus NS3 plays an essential role in virus assembly independent of its known enzymatic functions. J Virol 2008;82:3342–52 (PubMed: 18199634).

[65] Ramanathan MP, Chambers JA, Pankhong P, Chattergoon M, et al. Host cell killing by the West Nile Virus NS2B-NS3 proteolytic complex: NS3 alone is sufficient to recruit caspase-8-based apoptotic pathway. Virology 2006;345:56–72 (PubMed: 16243374).

[66] Miller S, Kastner S, Krijnse-Locker J, Buhler S, et al. The non-structural protein 4A of dengue virus is an integral membrane protein inducing membrane alterations in a 2K-regulated manner. J Biol Chem 2007;282:8873–82 (PubMed: 17276984).

[67] Roosendaal J, Westaway EG, Khromykh A, Mackenzie JM. Regulated cleavages at the West Nile virus NS4A-2K-NS4B junctions play a major role in rearranging cytoplasmic membranes and Golgi trafficking of the NS4A protein. J Virol 2006;80:4623–32 (PubMed: 16611922).

[68] Umareddy I, Chao A, Sampath A, Gu F, Vasudevan SG. Dengue virus NS4B interacts with NS3 and dissociates it from single-stranded RNA. J Gen Virol 2006;87:2605–14 (PubMed: 16894199).

[69] Munoz-Jordan JL, Laurent-Rolle M, Ashour J, et al. Inhibition of alpha/beta interferon signaling by the NS4B protein of flaviviruses. J Virol 2005;79:8004–13 (PubMed: 15956546).

[70] Ferrer-Orta C, Arias A, Escarmis C, Verdaguer N. A comparison of viral RNA-dependent RNA polymerases. Curr Opin Struct Biol 2006;16:27–34 (PubMed: 16364629).

[71] Yap TL, Xu T, Chen YL, et al. Crystal structure of the dengue virus RNA-dependent RNA polymerase catalytic domain at 1.85-angstrom resolution. J Virol 2007;81:4753–65 (PubMed: 17301146).

[72] Brooks AJ, Johansson M, John AV, et al. The interdomain region of dengue NS5 protein that binds to the viral helicase NS3 contains independently functional importin b1 and importin a/b-recognized nuclear localization signals (NLSs). J Biol Chem 2002;277:36399–407 (PubMed: 12105224).
[73] Pryor MJ, Rawlinson SM, Butcher RE, Barton CL, et al. Nuclear localization of dengue virus nonstructural protein 5 through its importin alpha/beta-recognized nuclear localization sequences is integral to viral infection. Traffic 2007;8:795–807 (PubMed: 17537211).
[74] Egloff MP, Benarroch D, Selisko B, Romette JL, et al. An RNA cap (nucleoside-2'-O)-methyltransferase in the flavivirus RNA polymerase NS5: crystal structure and functional characterization. Embo J 2002;21:2757–68 (PubMed: 12032088).
[75] Zhou Y, Ray D, Zhao Y, Dong H, Ren S, Li Z, et al. Structure and function of flavivirus NS5 methyltransferase. J Virol 2007;81:3891–903 (PubMed: 17267492).
[76] Filomatori CV, Lodeiro MF, Alvarez DE, Samsa MM, et al. A 5' RNA element promotes dengue virus RNA synthesis on a circular genome. Genes Dev 2006;20(16):2238–49.
[77] Alvarez DE, Filomatori CV, Gamarnik AV. Functional analysis of dengue virus cyclization sequences located at the 5' and 3' UTRs. Virology May 2008;375(1):223–35. https://doi.org/10.1016/j.virol.2008.01.014. Epub 2008 Mar 4.
[78] Wei Y, Qin C, Jiang T, et al. Translational regulation by the 3' untranslated region of the dengue type 2 virus genome. Am J Trop Med Hyg November 2009;81(5):817–24. https://doi.org/10.4269/ajtmh.2009.08.0595.
[79] Hahn CS, Hahn YS, Rice CM, Lee E, et al. Conserved elements in the 3' untranslated region of flavivirus RNAs and potential cyclization sequences. J Mol Biol 1987;198:33–41.
[80] Men R, Bray M, Clark D, Chanock RM, Lai CJ. Dengue type 4 virus mutants containing deletions in the 3' noncoding region of the RNA genome: analysis of growth restriction in cell culture and altered viremia pattern and immunogenicity in rhesus monkeys. J Virol 1996;70:3930–7.
[81] Leitmeyer KC, Vaughn DW, Watts DM, Salas R, et al. Dengue virus structural differences that correlate with pathogenesis. J Virol 1999;73:4738–47.
[82] Alvarez DE, De Lella Ezcurra AL, Fucito S, Gamarnik AV. Role of RNA structures present at the 3'UTR of dengue virus on translation, RNA synthesis, and viral replication. Virology September 2005;339(2):200–12.
[83] Olsthoorn RC, Mertens S, Brederode FT, Bol JF. A conformational switch at the 3' end of a plant virus RNA regulates viral replication. EMBO J 1999;18(17):4856–64.
[84] Pogany J, Fabian MR, White KA, Nagy PD. A replication silencer element in a plus-strand RNA virus. EMBO J 2003;22(20):5602–11.
[85] Zhang G, Zhang J, Simon AE. Repression and derepression of minus-strand synthesis in a plus-strand RNA virus replicon. J Virol 2004;78(14):7619–33.
[86] De Borba L, Villordo SM, Marsico FL, Carballeda JM, et al. RNA structure duplication in the dengue virus 3' UTR: redundancy or host specificity? mBio January 2019;10(1):e02506–18. https://doi.org/10.1128/mBio.02506-18.
[87] Villordo SM, Filomatori CV, Sanchez-Vargas I, Blair CD, Gamarnik AV. Dengue virus RNA structure specialization facilitates host adaptation. PLoS Pathog 2015;11:e1004604.
[88] Manzano M, Reichert ED, Polo S, et al. Identification of cis-acting elements in the 3'-untranslated region of the dengue virus type 2 RNA that modulate translation and replication. J Biol Chem 2011;286:22521–34.

[89] Manokaran G, Finol E, Wang C, Gunaratne J, et al. Dengue subgenomic RNA binds TRIM25 to inhibit interferon expression for epidemiological fitness. Science October 2015;350(6257):217–21. https://doi.org/10.1126/science.aab3369. Epub 2015 Jul 2.

[90] Göertz GP, Pijlman GP, et al. Dengue non-coding RNA: TRIMmed for transmission. Cell Host Microbe August 2015;18(2):133–4. https://doi.org/10.1016/j.chom.2015.07.009.

[91] Bidet K, Dadlani D, Garcia-Blanco MA, et al. G3BP1, G3BP2, and CAPRIN1 are required for translation of interferon stimulated mRNAs and are targeted by a dengue virus non-coding RNA. PLoS Pathog July 2014;10(7):e1004242. https://doi.org/10.1371/journal.-ppat.1004242. eCollection 2014 Jul.

[92] Perera N, Miller JL, Zitzmann N. The role of the unfolded protein response in dengue virus pathogenesis. Cell Microbiol May 2017;19(5). https://doi.org/10.1111/cmi.12734. Epub 2017 Mar 8.

[93] Diwaker D, Mishra KP, Ganju L. Effect of modulation of unfolded protein response pathway on dengue virus infection. Acta Biochimica Biophys Sin (Shanghai) 2015;47(12):960–8. https://doi.org/10.1093/abbs/gmv108.

[94] Smith JA. A new paradigm: innate immune sensing of viruses via the unfolded protein response. Front Microbiol 2014;5:222. https://doi.org/10.3389/fmicb.2014.00222.

[95] Celli J, Tsolis RM. Bacteria, the endoplasmic reticulum and the unfolded protein response: friends or foes? Nat Rev Microbiol 2015;13(2):71–82. https://doi.org/10.1038/nrmicro3393.

[96] Peña J, Harris E. Dengue virus modulates the unfolded protein response in a time-dependent manner. J Biol Chem 2011;286(16):14226–36. https://doi.org/10.1074/jbc.M111.222703.

[97] Jain B, Chaturvedi UC, Jain A. Role of intracellular events in the pathogenesis of dengue; an overview. Microb Pathog 2014;69–70:45–52. https://doi.org/10.1016/j.micpath.2014.03.004.

[98] Wu YW, Mettling C, Wu SR, Yu CY, et al. Autophagy-associated dengue vesicles promote viral transmission avoiding antibody neutralization. Sci Rep 2016;6:32243. https://doi.org/10.1038/srep32243.

[99] Heaton NS, Randall G. Dengue virus and autophagy. Viruses 2011;3:1332–41. https://doi.org/10.3390/v3081332.

[100] Green AM, Beatty PR, Hadjilaou A, Harris E. Innate immunity to dengue virus infection and subversion of antiviral responses. J Mol Biol 2014;426(6):1148–60. https://doi.org/10.1016/j.jmb.2013.11.023.

Chapter 7

# Clinical manifestations and laboratory diagnosis

Sargun Singh Walia[1,2], Mohammad A. Arif[3,4], Jahanzeb Liaqat[5]
[1]*Clinical Research Fellow, Zeenat Qureshi Stroke Institute, St. Cloud, MN, United States;* [2]*Department of Neurology, University of Missouri, Columbia, MO, United States;* [3]*Internal Medicine Shaheed Zulfiqar Ali Bhutto Medical University/Pakistan Institute of Medical Sciences, Islamabad, Punjab, Pakistan;* [4]*Internal Medicine, Ali Medical Centre, Islamabad, Punjab, Pakistan;* [5]*Lt. Colonel, Pak Emirates, Military Hospital, Rawalpindi, Pakistan*

## Clinical features of dengue viral illness

The full spectrum of illness and severity can be produced by any of the four dengue viral serotypes. In most cases, Dengue viral illness is self-limiting, however the febrile illness is often debilitating. The signs and symptoms can vary ranging from a mild, nonspecific febrile illness, to the more severe forms of the disease [1].

### Classification

In 1997, the World Health Organization classified Dengue viral illness was classified into three types on the basis of symptoms. It was divided into the following:

1. Dengue viral fever
2. Dengue viral hemorrhagic fever
3. Dengue viral shock syndrome

The initial case definitions of Dengue viral illness were given by World Health Organization in 1997. The classification was very complex and had varied applicability in the clinical setting. These definitions were not very efficient in guiding therapy for patients suffering from severe Dengue viral illness, but the disease did not qualifying to be categorized as Dengue viral hemorrhaging fever. A multicenter study was conducted in seven countries of Asia and Latin America in 2009 demonstrated gaps in existing case definition.

**Dengue Virus Disease. https://doi.org/10.1016/B978-0-12-818270-3.00007-2**

| World Health Organization classification 1997 | | |
|---|---|---|
| **1) Dengue viral fever** | **2) Dengue viral hemorrhagic fever** | **3) Dengue viral shock syndrome** |
| • Also called break bone fever.<br>• Criteria: Fever + two of the following -o retro-orbital pain or ocular pain<br>o Headache<br>o Muscle pain<br>o Bone/joint pain<br>o Rash<br>o Low white blood cell count<br>o Hemorrhagic manifestation: petechiae, positive tourniquet test, epistaxis, vaginal bleeding, hematuria, rectal bleeding, or hematemesis. | • Plasma leakage<br>• Hematocrit increase from baseline by $\geq$20%<br>• Dengue viral hemorrhagic fever criteria must include:<br>o Fever: biphasic lasting at least 2 −7 days<br>o Hemorrhagic manifestations including at least one of the following:<br>• Positive tourniquet test<br>• Bleeding from mucosa<br>• Melena or hematemesis<br>o Thrombocytopenia: Platelet levels $\leq$,100,000 cells/mm3<br>o Pleural effusion or ascites due to plasma leakage | • Dengue hemorrhagic fever leading to a circulating collapse due to plasma leakage<br>• Early sign—Narrowing pulse pressure<br>• Late sign—Hypotension<br>• Rapid and weak pulse<br>• Cold clammy skin |

The World Health Organization reclassified Dengue viral illness in 2009 on the basis of severity into:

**I.** Dengue viral illness without warning signs
**II.** Dengue viral illness with warning signs
**III.** Severe Dengue viral illness

According to World Health Organization, the prevalence of Dengue viral illness is increasing at an alarming rate. The reported case number increased from 2.2 million in 2010 to more than 3.34 million in 2016. Although, three World Health Organization member states regularly report the annual number of cases, global estimates of Dengue viral illness prevalence still remain uncertain. A study done to analyze the prevalence of Dengue viral illness cases in 128 countries has estimated that 3.9 billion people are at risk of Dengue viral illness [2]. Another study estimates that 390 million cases occur each year out of which 96 million show manifestation [3].

| World Health Organization classification 2009 | | |
|---|---|---|
| **1) Dengue viral illness without warning signs** | **2) Dengue viral illness with warning signs** | **3) Severe Dengue viral illness** |
| Fever + two of the following:<br>• Nausea, vomiting<br>• Rash<br>• Aches and pains<br>• Leukopenia<br>• Positive tourniquet test | Dengue viral illness fever with any of the following:<br>• Abdominal pain or tenderness<br>• Persistent vomiting<br>• Clinical fluid accumulation (ascites, pleural effusion)<br>• Mucosal bleeding<br>• Lethargy, restlessness<br>• Liver enlargement >2 cm<br>• Laboratory: Increase in HCT concurrent with rapid decrease in platelet count<br>• (Requires strict observation and medical intervention) | Dengue viral illness fever with atleast one of the following criteria<br>• Severe plasma leakage leading to:<br>a) Dengue shock syndrome<br>b) Fluid accumulation with respiratory distress<br>• Severe bleeding as evaluated by clinician<br>• Severe organ involvement<br>a) Liver: Aspartate transaminase and Alanine transaminase $\geq$1000<br>b) Neurologic: impaired consciousness<br>c) Failure of heart and other organs |

## Signs and symptoms

Adult patients with Dengue viral illness are more likely to present with clinical symptoms, whereas in children the infection is mostly asymptomatic. In endemic areas, history of prior infection with Dengue virus is very common. Secondary infection by a different serotype of Dengue virus can lead to more severe symptoms than the primary episode [3,4]. Clinical manifestation can start around day 4 after an infected mosquito bites a person. Dengue virus incubation period in humans ranges from 3 to 14 days [5].

The initial clinical manifestations of Dengue viral illness without warning signs and severe Dengue viral illness are similar, and the course of infection is short. Identifying patients that may develop severe Dengue viral illness is a challenging task. Severe Dengue viral illness may be distinguishable by clinical course that it passes through three stages of pathophysiology. These are the febrile phase, with high fever driven by viremia; the critical/plasma leak phase, which is manifested by sudden onset of varying degrees of plasma leak into the pleural and abdominal cavities; and convalescence/recovery/reabsorption phase, the hallmark of which is a sudden arrest of plasma leak with concomitant reabsorption of extravasated plasma and fluids Figs. 7.1 and

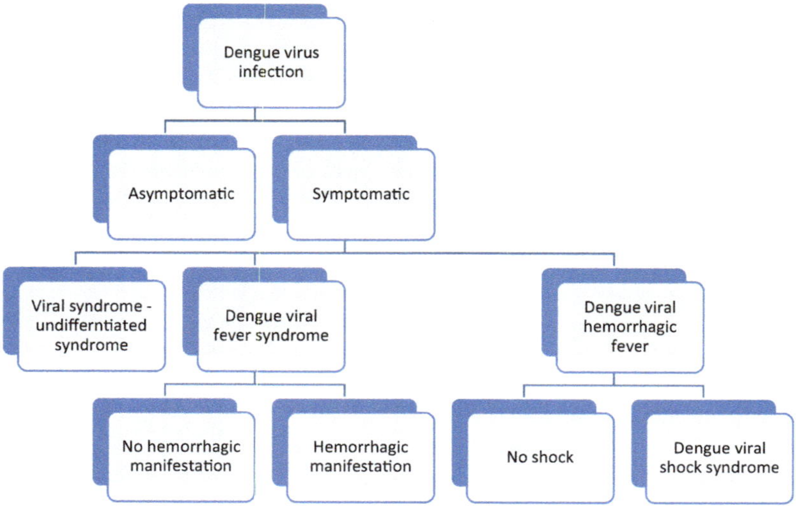


FIGURE 7.1 Adapted from Dengue hemorrhagic fever: diagnosis, treatment, prevention and control. World Health Organization 1997.

7.2. Figure 7.1 depicts the symptoms of Dengue viral illness divided into asymptomatic and symptomatic.

According to World Health Organization 1997 classification, Dengue hemorrhagic fever (DHF) and Dengue shock syndrome (DSS) includes all phases of infection i.e., febrile phase, critical phase, and the recovery phase. In Dengue fever (DF), there is no critical phase.

According to World Health Organization 2009 classification, severe dengue viral illness and dengue viral illness with warning signs include all phases of infection i.e., febrile phase, critical phase, and recovery phase. In dengue viral illness without warning signs, there is no critical phase. Figure 7.2 depicts the different phases of dengue viral infection classified by the World Health Organization in 1997 and reclassified in 2009.

## Febrile phase

Fever is one of the first symptoms to be seen in Dengue viral illness. The nonspecific nature of fever makes it challenging to distinguish from other febrile illnesses [6,7]. Other diseases that are mostly common in endemic areas of Dengue viral illness may include malaria, typhoid fever, and leptospirosis. In febrile phase, there is a sudden onset high grade fever ($\geq$101.3 F). The fever presents in a biphasic pattern in which the fever subsides and then reoccurs in 2 days. Fever can last anywhere from 2 to 7 days. Fever is associated with nonspecific symptoms including cranial, musculoskeletal, and gastrointestinal manifestations.

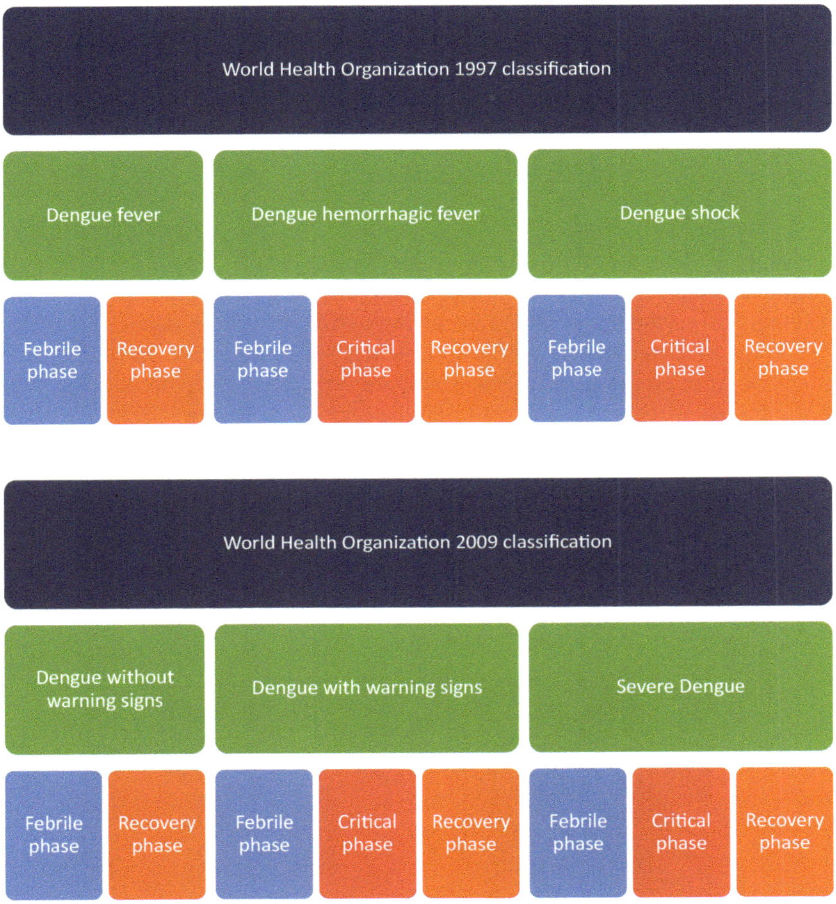


FIGURE 7.2 Different phases of infection in World Health Organization classifications of dengue viral illness from 1997 to 2009.

Severe headaches and retro-orbital eye pain are the major manifestations of cranial symptoms. Retro-orbital eye pain occurs due to various pathologies. The most accepted pathway is due to bleeding in the ophthalmic region secondary to thrombocytopenia. The bleeding in the macula and retinal periphery may lead to episodes of retro-orbital pain [8,9]. The patients top describe the pain as retro bulbar due to hemorrhage in the sub conjunctival region [10]. Others report the pain to be diffuse.

The musculoskeletal features include arthralgia and myalgia [11]. Dengue viral illness is also known as the break bone fever. Diffuse muscle and joint pain is one of the key symptoms of dengue viral illness. It is also known as Dengue associated muscle dysfunction (DAMD). The pathogenesis behind these pains is believed to be due to interstitial hemorrhage, necrosis, and edema. Occasionally myophagocytosis has also been seen in patients suffering from Dengue viral illness. Most of the patients improve within 7–14 days of infection [12]. Serum creatinine kinase levels may be raised due to muscle breakdown. In rare cases, mortality due to cardiomyopathy has also been reported [13,14]. Gastrointestinal symptoms can occur varying from intermittent nausea and vomiting to anorexia.

A characteristic rash occurs in majority of patients suffering from fever secondary to Dengue viral illness. Primary rash occurs within 1–2 days of symptom onset and typically starts from the face. The rash is transient, flushing, and erythemic in nature occurring due to dilation of capillaries. After 3–6 days of fever onset, a transient second rash appears. Mobiliform or maculopapular eruptions are seen which are typically asymptomatic Fig. 7.3. In a minority of patients, this rash may be pruritic in nature [15,16].

Hemorrhagic manifestations such as petechiae and bleeding from mucosal membrane may be seen. Minor, subtle petechial hemorrhages are often found on the lower extremities, but may also be encountered on the hard and soft palates, buccal mucosa, and the subconjunctivae. Petechial rash generally starts from lower extremities and then spreads to the thorax and other body parts. In some rare cases, vaginal hemorrhage has been reported in pregnant

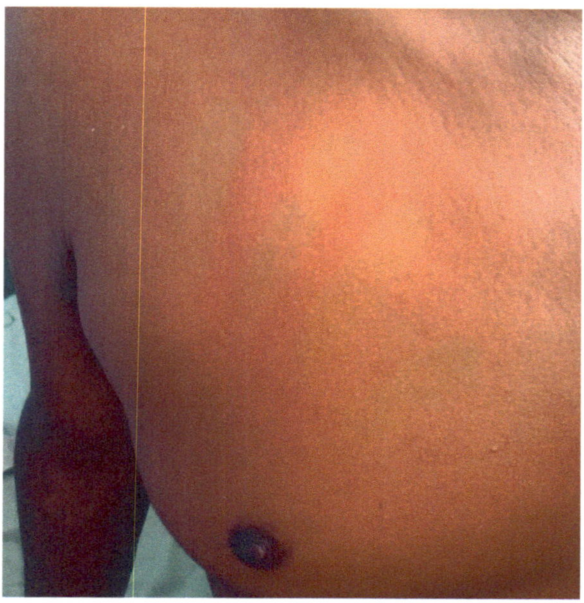

FIGURE 7.3 Early dengue fever rash. *Source: commons.wikimedia.org. Attribution: Ranjan Premaratna. Professor in medicine. Department of medicine. University of Kelaniya. Sri Lanka [CC BY-SA 4.0].*

women. A positive tourniquet test can be used as to test the fragility of capillaries [17]. In a tourniquet test, a blood pressure cuff is placed around the upper arm. The tourniquet is kept inflated for 5 min up to the middle pressure between systolic and diastolic pressures. A positive test is identified by more than 20 petechiae per 2.5 cm2 area Figs. 7.4–7.7.

These clinical features do not predict the severity of Dengue viral illness. Therefore, it is crucial to monitor for warning signs and other clinical parameters in order to recognize progression to the critical phase. Dengue viral illness fever is often misdiagnosed as influenza or other viral diseases in the absence of a rash especially if the other symptoms are mild.

## Critical phase

Patients with subsidence of fever will improve and not transition into the critical phase. Patients that do transition into the critical phase present with warning signs and symptoms of severe Dengue viral illness [18]. Gastrointestinal symptoms worsen in the form of persistent vomiting and severe abdominal pain. Bleeding can be seen from previous venipuncture sites. Dyspnea and lethargy is seen in majority of cases. Patients with increased capillary permeability transition from febrile phase into the critical phase. Critical phase lasts between 24 and 48 h. This phase will present with symptoms indicative of plasma leakage. This can manifest as a pulmonary syndrome and/or renal syndrome [19]. The patients present with worsening of symptoms which typically occurs around day 3–7. Young adults and children develop a systemic vascular leak syndrome. This typically presents as shock, bleeding episodes, plasma leakage, and organ impairment or failure. Initially it presents with a hypotensive phase with a significant decrease in blood pressure. The hypotension is accompanied by hemoconcentration [20,21]. Capillary leak can be first evidenced with the presence of increased hematocrit but a decrease in albumin levels [22]. If left untreated, the disease can lead to 20% mortality rate. Proper management with intravenous hydration can reduce this mortality rate to less than 1% [23]. Systemic vascular leak syndrome is preceded by progressive leukopenia ($\leq$5000 cells/mm3) with a rapid decline in platelet count to about 100,000 cells/mm3. Rise in hematocrit is one of the earliest signs of plasma leakage and can reflect the severity of plasma leakage. High volume plasma leak can result in pleural effusion causing increasing respiratory distress, ascites, gastrointestinal bleeding, and hypovolemic shock. Hypoperfusion of tissues can cause metabolic acidosis, progressive organ failure, and disseminated intravascular hypoperfusion. Disseminated intravascular hypoperfusion can result in severe hemorrhage leading to a decrease in hematocrit level. White cell count is often increased in patients with severe hemorrhage. Acute kidney injury can rarely occur in Dengue viral illness. Hematuria and proteinuria can be seen along with glomerular abnormalities [24–26]. Initially the patient might appear normal, but immediate fluid

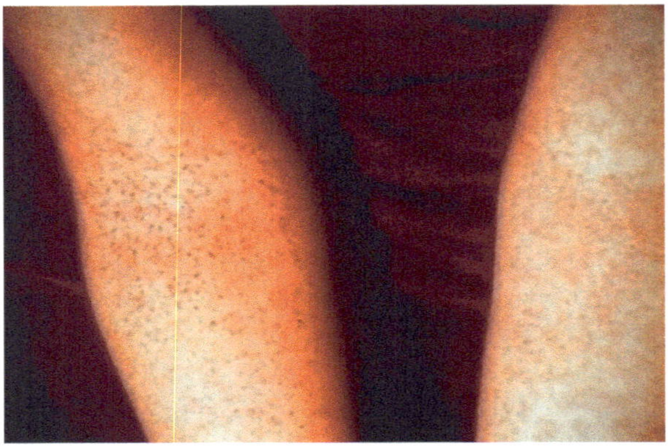

FIGURE 7.4 A positive tourniquet test on the left side of the image in a person with dengue fever. *Source: https://commons.wikimedia.org/wiki/File:Positive-tourniquet-test.gif; Center of disease control and prevention. Attribution: Center for disease control and prevention [Public domain].*

resuscitation is necessary to prevent the complications of fluid leakage. The initial increase in hematocrit levels go back to normal or subnormal levels after fluid administration. Frequent hematocrit level checks are used to guide intravenous fluid administration to counter the plasma leak [27]. Plasma leak can also be seen in Ebola virus, Marburg virus, and Hantavirus illnesses.

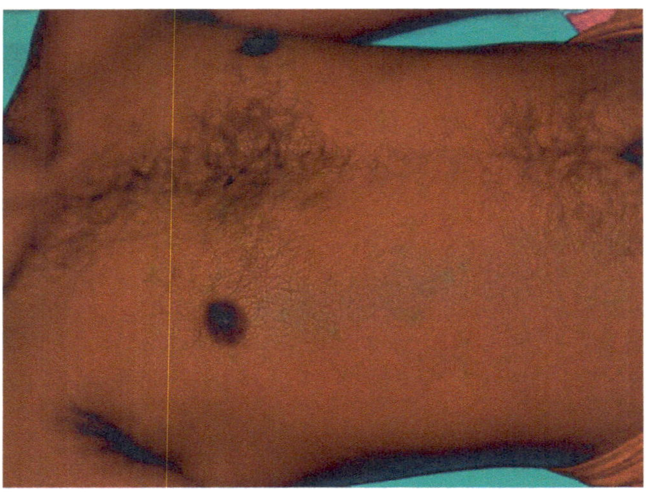

FIGURE 7.5 "Isles of white in sea of red." *Source: https://commons.wikimedia.org/wiki/File:Dengue_recovery_rash_(White_islands_in_red_sea).jpg Attribution: Ranjan Premaratna. Professor in medicine. Department of medicine. University of kelaniya. Sri Lanka. [CC BY-SA 4.0].*

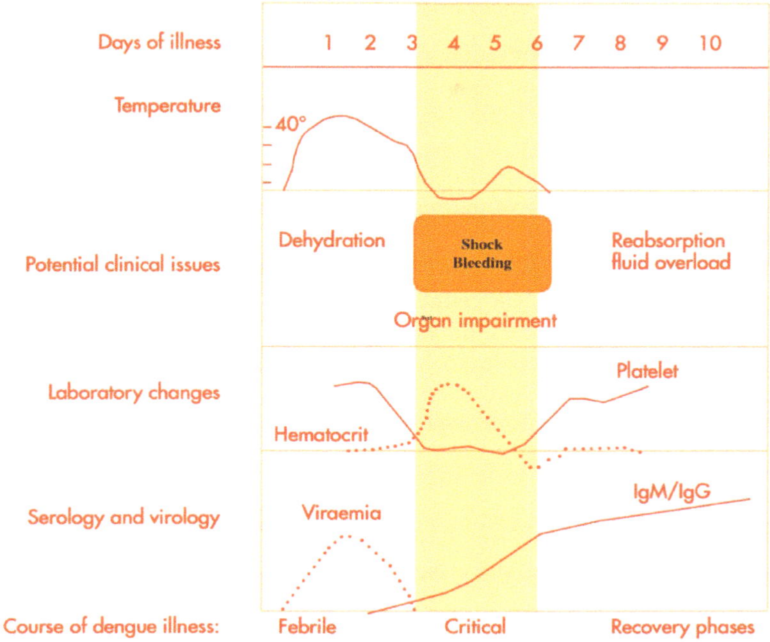


**FIGURE 7.6** Course of Dengue viral illness. *Source: Adapted from Dengue: guidelines for diagnosis, treatment, prevention and control. © World Health Organization 2009.*

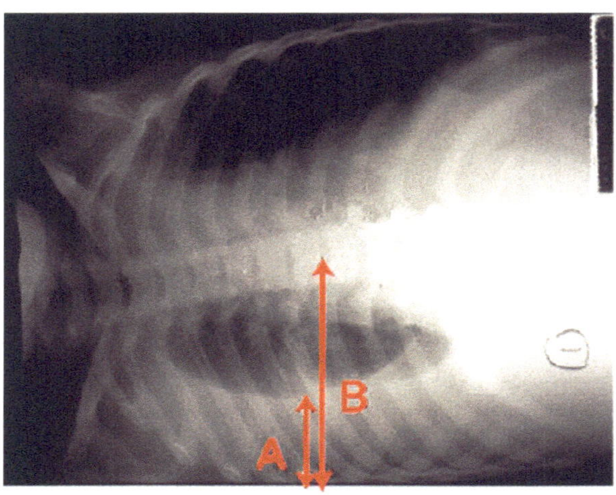


**FIGURE 7.7** Chest X-ray showing pleural effusion. Arrow A shows fluid accumulation and arrow B shows the normal width of the lung. *Centers for disease control and prevention.*

## Recovery phase

After the critical phase subsides, slow reabsorption of fluid starts from the extravascular compartment. This takes approximately 48–72 h. During this phase hemodynamic status improves, vital signs stabilize and gastrointestinal symptoms start improving. In rare cases, an erythematous or petechial rash may be seen. There are small patches of normal skin in between the rash and is referred as "isles of white in the sea of red" Figure 7.5.

Electrocardiographic changes and bradycardia are common during this stage. Hematocrit decreases to normal levels or a little below baseline due to hemodilution. The leukocyte count starts rising after fever subsides. Use of intravenous fluid should be cautious as excessive use can lead to complications such as respiratory distress from massive pleural effusion and ascites, pulmonary edema, or congestive heart failure. Figure 7.6 depicts the course of Dengue viral illness in various phases of infection.

## Complications

Various complications can arise during the different phases of infection, which should be carefully screened for include the following:

1) Febrile phase:
   a) Dehydration
   b) Neurological symptoms: encephalopathy and seizures
2) Critical phase:
   a) Shock
   b) Hemorrhage
   c) Organ impairment: Acute kidney injury [28].
3) Recovery phase
   a) Excessive intravenous fluid administration leading to hypervolemia
   b) Acute pulmonary edema

## Diagnosis

Since there is no therapeutic agent available for Dengue viral illness treatment, successful management depends upon timely and judicious use of supportive care, including administration of isotonic intravenous fluids or colloids, and close monitoring of vital signs and hemodynamic status, fluid balance, and hematologic parameters. Dengue viral illness may be routinely diagnosed based on clinical manifestation. Laboratory test can help in diagnosing a Dengue viral illness. These tests can be broadly divided into nonspecific tests and definitive tests. Example of a nonspecific test includes a complete blood count panel, while an example of a definitive test is a nonstructural protein 1 (NS1) antigen assay and serology testing. A study done in 2011 published that approximately 50% of primary care physicians use definitive tests to diagnose Dengue viral illness [29].

Use of diagnostic techniques vary based on the stage of infection. Basic laboratory test can help in the detection and management of Dengue viral illness. In the early stages of the infection, isolation of virus, nucleic acid, or antigen serves as the best method of detection. In the later stage or at the end of the acute phase of infection, routine laboratory diagnostics and a clinical examination often do not lead to a definitive diagnosis. The diagnosis of Dengue viral illness will remain elusive unless serological or molecular tests for identification of Dengue virus are undertaken. The diagnosis is confirmed by serological tests for antidengue virus IgM antibodies or dengue virus ribonucleic acid by molecular testing.

Definite diagnosis of Dengue viral illness requires isolation of virus but is not practical as it takes several weeks for results to be available [30]. If less than 5 days have passed since the onset of fever then definitive and serotype specific identification tests are used. For rapid diagnosis, viral nucleic acids are detected by reverse transcriptase-polymerase chain reaction and secreted a nonstructural protein 1 antigen is capture by enzyme-linked immunosorbent assays [31]. These tests can help in managing cases especially in primary care setting due to clinical uncertainties. Nonstructural protein 1 antigen testing can be done at a very lower cost and recommended by the Ministry of Health in Singapore to be done in less than 7 days from symptom onset [32]. Nonstructural protein 1 antigen detection is a highly sensitive and specific test for Dengue viral illness [33,34]. Sensitivity is of a nonstructural protein 1 has been found to be as high as 90% in primary dengue viral illness and 60%−80% in secondary Dengue viral illness [35]. Reverse transcriptase-polymerase chain reaction becomes positive within 5 days of illness and is very sensitive and specific if performed in laboratory with specialized equipment and trained staff. In clinical practice most of the tests are now done using commercially available kits which are less reliable due to the lack of standardization and quality control [36].

Nonstructural protein 1 viral protein is secreted from infected cells, which can be detected in blood of infected individual within 7 days of illness [37]. The level of a nonstructural protein 1 seems to correlate with viral titers and can be viewed as a surrogate marker for viremia [38,39]. Now most of the hospitals are using a nonstructural protein 1 capture by enzyme-linked immunosorbent assay and rapid strip test for early diagnosis of Dengue viral illness especially in primary infection. In secondary infection, preformed antibodies against nonstructural protein 1 antigen sequester nonstructural protein 1 in immune complexes thus interfering with detection by assay.

Serological diagnosis of Dengue viral illness can be made by detection of IgM antibodies as early as four days after the onset of illness. In a patient presenting with clinical features of Dengue viral illness, presence of IgM antibodies in a single serum sample has been widely used to get a preliminary diagnosis. The time for the patient to form antibodies against the Dengue viral illness is called seroconversion. The seroconversion is seen in seven days in primary infection and in four days in secondary infection. Definite diagnosis requires seroconversion and a fourfold rise in titers of antibodies between paired and convalescent phase samples taken 2 weeks apart [40]. A hemagglutination inhibition assay antibody titer of 1280 or higher is diagnostic of a probable dengue viral illness. Both probable and confirmed dengue viral illness cases should be reported to health authorities. Serological test are not reliable in patients who have been vaccinated [41] or had recent infection with antigenically related viruses like yellow fever virus, Japanese encephalitis virus or zika virus. Viral proteins are detected by immunohistochemical staining of tissue samples [42]. Liver biopsy has the highest yield in this regard and is usually done for postmortem diagnosis of a suspected case of Dengue viral illness.

## Viral isolation

Specimens are usually collected before day 5 in the early phase of infection in the period of viremia [43]. Samples are collected from serum, peripheral blood, or biopsy tissue. Most commonly used method to isolate Dengue virus is cell culture. Host cells used are mosquito cell AP61 (cell line from Ae. pseudoscutellaris) or line C6/36 (cloned from Ae. albopictus) [44–46]. The specimen should be properly stored and transported to preserve the viability [47]. It takes 1–2 weeks to isolate the virus and confirm using immunofluorescence assay. Suckling mice can also serve as clinical specimens and are used to inoculate via intracranial route. Virus antigens can then be detected using anti-Dengue antibody staining in mouse brain samples.

## Nucleic acid detection

Ribonucleic acid (RNA) is very unstable and heat labile and requires proper handling. Similar methods are used for storage and transport as used in nucleic acid detection methods. Reverse transcriptase-polymerase chain reactions have been used since 1990s to detect nucleic acids [48]. This offers better sensitivity (80%–100%) than viral isolation methods and the detection time is also lesser. Principal steps involved are the following: nucleic acid extraction and purification; amplification of the nucleic acid; and detection and characterization of the amplified product. Real time reverse transcriptase-polymerase chain reaction is a one-step system that can be used to detect and quantify viral RNA specific to different Dengue serotypes [49]. This test comes in two types of kits that are either singleplex type or multiplex type [50]. A singleplex type kit can detect one serotype at a time whereas a multiplex type kit can detect all four serotypes in a single sample. Other methods used for nucleic acid detection include nucleic acid sequence–based amplification assay [51–53]. Nucleic acid sequence–based amplification assay has been adapted to be used to study Dengue viral illness in the field. Loop mediated amplification method is another way to detect nucleic acid [54–56].

## Antigen detection

Antigen detection in patients suffering from secondary infection is difficult in an acute-phase serum sample as these patients have preexisting virus Immunoglobulin G antibodies. However with the developments in enzyme-linked immunosorbent assay and dot blot assays such diagnosis has been made easy. High concentrations of nonstructural protein 1 and envelop/membrane antigen immune complexes can be detected in both primary and secondary dengue viral illness. This can be done within 9 days after the onset of symptoms. Nonstructural protein 1 kits are available commercially which can help in early diagnosis of dengue viral illness [57,58]. But such kits cannot differentiate between different serotypes of infection. Immunoperoxidase, avidin–biotin enzyme, and fluorescent antibody assays can be used in autopsy biopsy samples.

## Immunoglobulin M antibody-capture enzyme-linked immunosorbent assay (MAC-ELISA)

Immunoglobulin M antibody-capture enzyme-linked immunosorbent assay (MAC-ELISA) is a technique to test antibodies against Dengue virus [59]. A microplate is coated with anti-μ chain–specific antibodies which then captures total Immunoglobulin M in the serum sample. Antigens specific to one of the serotypes of Dengue virus attach to the captured dengue virus antibodies. These are then detected via directly or indirectly conjugated enzyme which transforms into colored substrate. Spectrophotometer is used to measure optical density. Serum samples are collected after 5 days or more from symptom onset. Immunoglobulin M antibody-capture enzyme-linked immunosorbent assay is highly sensitive and specific if used after 5 days of symptom onset.

Rapid commercial kits in the form of enzyme-linked immunosorbent assay are available [60]. Cross reactivity has been reported with malaria and secondary Dengue infection [61]. The disadvantage of Immunoglobulin M antibody-capture enzyme-linked immunosorbent assay is that it cannot differentiate among different serotypes of Dengue virus [62].

## Immunoglobulin G of enzyme-linked immunosorbent assay (Immunoglobulin G ELISA)

Immunoglobulin G enzyme-linked immunosorbent assay uses the same type of antigens used in Immunoglobulin M antibody-capture enzyme-linked immunosorbent assay. The advantages of using Immunoglobulin G enzyme-linked immunosorbent assay is that it can detect both recent and past Dengue viral illness. It can even detect Immunoglobulin G antibodies after 10 months of Dengue viral illness using E/M-specific capture Immunoglobulin G enzyme-linked immunosorbent assay (GAC). Immunoglobulin G antibodies can be detected for life but a 4x increase in number can be used to detect a recent infection. This test can also be used for the surveillance and serological diagnosis of Dengue viral illness cases. This can also be used to differentiate between a primary and a secondary infection.

## Immunoglobulin M/Immunoglobulin G ratio

Immunoglobulin M/Immunoglobulin G can be used to distinguish primary from secondary dengue viral illness. If the immunoglobulin M/immunoglobulin G optical density ratio is $\geq 1.2$ or 1.4 (depending on the dilution level), it is termed as primary infection. If immunoglobulin M/immunoglobulin G optical density ratio is $\leq 1.2$ or 1.4 the dengue viral illness is classified as a secondary infection.

## Immunoglobulin A

Antidengue virus immunoglobulin A capture enzyme-linked immunosorbent assay is used to detect antidengue immunoglobulin A antibodies in the patient sera. It usually becomes positive 1 day after Immunoglobulin M. Peak levels can be recorded around day 8 of symptom onset.

## Hemagglutination–inhibition test

Principal of this test is that Dengue virus antigens agglutinate red blood cells (RBCs). This agglutination process is inhibited by anti-Dengue antibodies which can be measured in a hemagglutination–inhibition test. Paired serum samples should be collected on admission and on discharge or the samples should be spaced 7 days. In a primary dengue viral illness low level of antibodies in samples taken before 5 days which starts elevating slowly thereafter. In secondary infection, hemagglutination–inhibition antibody titer levels rise at a much rapid rate exceeding 1:1280.

## Other laboratory diagnostic modalities

Dengue fever is suspected in endemic areas when a patient with acute febrile illness develops thrombocytopenia and bleeding complications. Other diseases that can lead to thrombocytopenia besides Dengue viral illness are malaria, rickettsial infections, scrub typhus, leptospira, and meningococci infections. The decrease in platelet count occurs due to reduction in platelet production and increased destruction of platelets [63]. Sepsis by gram positive and gram negative bacterial infections can lead to thrombocytopenia. Platelet specific immunoglobulin G antibodies can cause the platelets to attach to damaged vascular surfaces. World Health Organization suggests using tourniquet test for early diagnosis of a suspected case of dengue fever [64] as definite diagnosis in acute settings has little impact in overall management of the patient.

Basic laboratory tests that are routinely used in the management of suspected Dengue viral illness cases are as follows:

1) Complete blood count [65,66].
   - Leukopenia: White cell count less than equal to 5000 cells/mm3 in the early stages of infection. It later normalizes after defervescence. Lymphocytosis may be seen in the presence of shock.
   - Thrombocytopenia: Platelet count is less than 100,000 cells/mm3. Platelet count recovery is typically slow even in recovery stages of infection. Platelet levels should be reassessed every 24 h to detect conversion to Dengue hemorrhagic fever.
   - Hematocrit increase by more than equal to 20% above the baseline. This may vary with the level of fluid administration (lower due to dilutional effect) and hemorrhage from the gastrointestinal tract. A study done in Thailand in 2004 found that hematocrit levels done at the time of admission cannot predict the future outcome of Dengue hemorrhagic fever and dengue shock syndrome [67]. It emphasized that repeated levels are necessary to better predict outcome. Hematocrit levels should be rechecked every 24 h to detect dengue hemorrhagic fever in the early stages. If Dengue shock syndrome is suspected then the hematocrit levels should be repeated every 3–4 h.

2) Metabolic panel
   - Hypoproteinemia
   - Metabolic acidosis

- Electrolyte disturbance: Hyponatremia may be seen in dengue hemorrhagic fever or dengue shock syndrome.
- Blood urea nitrogen (BUN): Levels may be increased in shock.

3) Liver panel [65,68].
   - Serum aspartate transaminase (AST) is usually elevated to 2 to 5 times the upper limits but marked elevation up to 15 times can also occur.
   - Jaundice and acute liver failure are uncommon due to direct viral effect but have been described in dengue shock syndrome. Prolonged hypoperfusion and hypoxia are presumed to be the contributory factors [69].

4) Coagulation profile and disseminated intravascular coagulation (DIC) panel.
   - Prolonged prothrombin time
   - Prolonged activated partial thromboplastin time
   - Decreased fibrinogen
   - Increased amount of fibrin split products
   - Signs of early coagulopathy may be very subtle.

5) Guaiac test
   - Should be used in all patients suspected with Dengue viral illness.
   - Positive for occult blood in the stool.

6) Urinalysis
   - Used to identify hematuria

7) Cultures
   - Blood, urine and, cerebrospinal fluid cultures may be performed to exclude other pathological causes.

8) Arterial blood gas

Other than the basic information, medical professionals should also record travel history and history of onset of symptoms. Capillary leak syndrome can cause development of bodily effusions, which can be detected as early as 3 days after the infection with the help of ultrasound of chest and abdomen [27]. Right lateral decubitus chest X-ray is useful for detection of development of pleural effusion in places where ultrasound facility is not available Figure 7.7. In most centers, imaging is done on daily basis after the first few days of infection so that capillary leak is detected very early on and appropriate management can be instituted Table 7.1.

**TABLE 7.1 Summary of Dengue viral illness diagnostic methods.**

| Dengue viral illness | Detection method—ELISA | Interpretation | Sample collection | Time of collection after onset of symptoms |
|---|---|---|---|---|
| Confirmed | Virus | Virus isolated | • Serum (day 1–5)<br>• Necropsy tissue | 1–2 weeks |
| | Genome detection | + ve RT-PCR or + ve real time RT-PCR | | 1 or 2 days |
| | Antigen detection | + ve NS1 antigen | | >1 day |
| | | + Immunohistochemistry | • Necropsy tissue | |
| | Immunoglobulin M seroconversion | Conversion from −ve Immunoglobulin M to + ve Immunoglobulin M in paired serum samples | • Serum (day 1–5)<br>• Convalescent serum (15–21 days after 1st sample) | Rapid test 30 min/ELISA 1–2 days |
| | Immunoglobulin G seroconversion | Conversion from −ve Immunoglobulin G to + ve Immunoglobulin G in paired serum samples/4× increase in Immunoglobulin G levels in paired samples | | 7 days or more |
| Probable | Immunoglobulin M | +ve titers | • Serum after day 5 | 1–2 days |
| | Immunoglobulin G | High serum Immunoglobulin G titers on hemagglutination inhibition assay (≥1280) | | 7 days or more |

Adapted from Dengue and Control (DENCO) Study and.ELISA (Enzyme-linked immune sorbent assay); NS1 Antigen (nonstructural protein 1 antigen); Immunoglobulin M; Immunoglobulin G; RT-PCR (reverse transcriptase-polymerase chain reaction).

## Future diagnostics methods under development

### *Microsphere-based immunoassays (MIA)*

It is a highly advanced serology diagnostic technique which can be used for laboratory testing for variety of diseases [70]. In this, antibody or antigens are covalently bonded to microspheres or beads. With the help of lasers, fluorescence of varying wavelengths can be elicited. One of the advantages of microsphere-based immunoassays over IgM antibody capture enzyme-linked immunosorbent assay is faster diagnostic time [71]. It also has the ability to identify antibody response to different viruses by multiplexing serological tests [72,73].

### *Biosensor technology*

This technology is developing very rapidly [74]. It uses mass spectrometry and is helpful in discriminating biological components in a complex mixture. It produces mass spectra and serves as a fingerprint of the bacteria or virus making a molecular profile. The software used by the device has a vast database [75]. It uses this database to compare the mass spectra produced to identify the pathogen. It can also identify a pathogen which does not exist in the database by comparing the similarities with the other pathogens. This test can be very helpful in identifying different serotypes of Dengue virus in an outbreak. Identification kits are being developed to take samples which can later be processed for DNA, mass spectrometry, and identification analysis [76].

### *Microarray technology*

This test is an advanced diagnostic tool that can screen a sample for multiple viruses by analyzing their nucleic acid fragments [77]. The sample needs to be amplified and then hybridization is done on a microarray. This can analyze both random sequences and conserved sequence [78]. The advantage of this technology is that it can detect divergent strains of viruses. The DNA fragments are labeled with fluorescent dyes. This test is then analyzed using a laser based scanner. It can be used to test patients in a region which is endemic for dengue viral illness and other arboviruses than can mimic symptoms of a dengue viral illness [79].

### *Luminescence technology*

The test is still under early development. Advantage of this system will be its inexpensive instruments.

## Differential diagnosis

Many different etiologies can present in a similar way as Dengue viral illness. Some of the infection that may present in a similar way as Dengue viral illness are as follows:

## Chikungunya virus infection

Chikungunya and Dengue are both transmitted by *Aedes aegypti* and *Aedes albopictus* mosquitoes. On comparison both may present with a febrile illness along with a rash. Both lead to arthralgias but in chikungunya joint swelling (inflammatory arthritis) is a hallmark symptom whereas thrombocytopenia and bleeding symptoms are more specific to dengue [80]. These can also be differentiated using reverse transcriptase-polymerase chain reaction (RT-PCR) [81].

## Zika virus

Zika and Dengue viral illness are both transmitted by *Aedes aegypti* and *Aedes albopictus* mosquitoes. Zika virus infection leads to conjunctivitis in patients invariably, which is not seen in dengue viral illness. Reverse transcriptase-polymerase chain reaction can be used to definitely differentiate between them. In rare cases, a coinfection between Dengue, Zika, and Chikungunya has been reported [82].

## Typhoid

Both typhoid and dengue may present with fever, abdominal pain, and rash. An easy way to distinguish them is by using blood or stool culture. Raman spectroscopy has also been used to differentiate between them [83].

## Rickettsial infection

Also known as African tick bite fever, is most commonly seen in people traveling to Africa and Caribbean islands. The infection presents with similar symptoms as dengue that include fever, headache, and myalgias. The differentiating features include the presence of solitary or multiple eschars that are associated with regional lymphadenopathy. It can be differentiated from dengue with the use of direct smear and the use of Polymerase Chain Reaction techniques [84].

## Malaria

Malaria infection can be differentiated from dngue by visualizing parasites in peripheral blood smears [85].

## Hemorrhagic fever

Multiple viruses can lead to hemorrhagic fever like Dengue viral illness. These include Ebola virus, Lassa virus, Hantavirus, Crimean-congo virus, Yellow fever virus, and Marburg virus. The easiest way to distinguish among them is by epidemiological exposure. Confirmatory tests include polymerase chain reaction and viral serology.

## Bacterial sepsis

Bacterial sepsis may also present with fever and even shock. Blood cultures are an easy way to distinguish it from dengue viral illness.

## References

[1] Whitehorn J, Simmons CP. The pathogenesis of dengue. Vaccine 2011;29(42):7221–8.
[2] Brady OJ, et al. Refining the global spatial limits of dengue virus transmission by evidence-based consensus. PLoS Negl Trop Dis 2012;6(8):e1760.
[3] Bhatt S, et al. The global distribution and burden of dengue. Nature 2013;496(7446):504–7.
[4] Bouzid M, et al. Climate change and the emergence of vector-borne diseases in Europe: case study of dengue fever. BMC Public Health 2014;14:781.
[5] Fontaine A, et al. Epidemiological significance of dengue virus genetic variation in mosquito infection dynamics. PLoS Pathog 2018;14(7):e1007187.
[6] Zavala-Velazquez JE, Yu XJ, Walker DH. Unrecognized spotted fever group rickettsiosis masquerading as dengue fever in Mexico. Am J Trop Med Hyg 1996;55(2):157–9.
[7] Watt G, et al. Differentiating dengue virus infection from scrub typhus in Thai adults with fever. Am J Trop Med Hyg 2003;68(5):536–8.
[8] Lim WK, et al. Ocular manifestations of dengue fever. Ophthalmology 2004;111(11):2057–64.
[9] Chan DPL, et al. Ophthalmic complications of dengue. Emerg Infect Dis 2006;12(2):285–9.
[10] Kapoor HK, et al. Ocular manifestations of dengue fever in an East Indian epidemic. Can J Ophthalmol 2006;41(6):741–6.
[11] Bhaskar E, Sowmya G, Moorthy S. Musculoskeletal manifestations of dengue fever: is there a changing pattern? J Clin Rheumatol 2012;18(2):102–3.
[12] Misra UK, et al. Dengue-associated transient muscle dysfunction: clinical, electromyography and histopathological changes. Infection 2012;40(2):125–30.
[13] Salgado DM, et al. Heart and skeletal muscle are targets of dengue virus infection. Pediatr Infect Dis J 2010;29(3):238–42.
[14] Wali JP, et al. Cardiac involvement in dengue haemorrhagic fever. Int J Cardiol 1998;64(1):31–6.
[15] Chadwick D, et al. Distinguishing dengue fever from other infections on the basis of simple clinical and laboratory features: application of logistic regression analysis. J Clin Virol 2006;35(2):147–53.
[16] Thomas EA, John M, Bhatia A. Cutaneous manifestations of dengue viral infection in Punjab (north India). Int J Dermatol 2007;46(7):715–9.
[17] Halsey ES, et al. Performance of the tourniquet test for diagnosing dengue in Peru. Am J Trop Med Hyg 2013;89(1):99–104.
[18] Kalayanarooj S. Clinical manifestations and management of dengue/DHF/DSS. Trop Med Health 2011;39(4 Suppl. 1):83–7.
[19] Peters CJ, Simpson GL, Levy H. Spectrum of hantavirus infection: hemorrhagic fever with renal syndrome and hantavirus pulmonary syndrome. Annu Rev Med 1999;50:531–45.
[20] Earle DP. Analysis of sequential physiologic derangements in epidemic hemorrhagic fever; with a commentary on management. Am J Med 1954;16(5):690–709.
[21] Organization WH, et al. Dengue: guidelines for diagnosis, treatment, prevention and control. World Health Organization; 2009.

[22] Giles RB, Langdon EA. Blood volume in epidemic hemorrhagic fever. Am J Med 1954;16(5):654–61.
[23] Garcia G, et al. Long-term persistence of clinical symptoms in dengue-infected persons and its association with immunological disorders. Int J Infect Dis 2011;15(1):e38–43.
[24] Sheedy JA, et al. The clinical course of epidemic hemorrhagic fever. Am J Med 1954;16(5):619–28.
[25] Lima EQ, Nogueira ML. Viral hemorrhagic fever-induced acute kidney injury. Semin Nephrol 2008;28(4):409–15.
[26] van Ypersele de Strihou C. Clinical features of hemorrhagic fever with renal syndrome in Europe. Kidney Int Suppl 1991;35:S80–3.
[27] Srikiatkhachorn A, et al. Natural history of plasma leakage in dengue hemorrhagic fever: a serial ultrasonographic study. Pediatr Infect Dis J 2007;26(4):283–90. discussion 291-2.
[28] Lombardi R, et al. Epidemiology of acute kidney injury in Latin America. Semin Nephrol 2008;28(4):320–9.
[29] Lee LK, et al. Dengue knowledge, attitudes, and practices among primary care physicians in Singapore. Ann Acad Med Singapore 2011;40(12):533–8.
[30] Lanciotti RS, et al. Rapid detection and typing of dengue viruses from clinical samples by using reverse transcriptase-polymerase chain reaction. J Clin Microbiol 1992;30(3):545–51.
[31] Shu PY, et al. Dengue virus serotyping based on envelope and membrane and nonstructural protein NS1 serotype-specific capture immunoglobulin M enzyme-linked immunosorbent assays. J Clin Microbiol 2004;42(6):2489–94.
[32] Wai MNC. Revisiting the approach to dengue: the primary care perspective. Singap Fam Physician 2015;41(2):65–73.
[33] Solanke VN, Karmarkar MG, Mehta PR. Early dengue diagnosis: role of rapid NS1 antigen, NS1 early ELISA, and PCR assay. Trop J Med Res 2015;18(2):95.
[34] Chaterji S, et al. Evaluation of the NS1 rapid test and the WHO dengue classification schemes for use as bedside diagnosis of acute dengue fever in adults. Am J Trop Med Hyg 2011;84(2):224–8.
[35] Hang VT, et al. Diagnostic accuracy of NS1 ELISA and lateral flow rapid tests for dengue sensitivity, specificity and relationship to viraemia and antibody responses. PLoS Negl Trop Dis 2009;3(1):e360.
[36] Najioullah F, Viron F, Cesaire R. Evaluation of four commercial real-time RT-PCR kits for the detection of dengue viruses in clinical samples. Virol J 2014;11:164.
[37] Guzman MG, et al. Multi-country evaluation of the sensitivity and specificity of two commercially-available NS1 ELISA assays for Dengue diagnosis. PLoS Negl Trop Dis 2010;4(8).
[38] Young PR, et al. An antigen capture enzyme-linked immunosorbent assay reveals high levels of the dengue virus protein NS1 in the sera of infected patients. J Clin Microbiol 2000;38(3):1053–7.
[39] Libraty DH, et al. Differing influences of virus burden and immune activation on disease severity in secondary dengue-3 virus infections. J Infect Dis 2002;185(9):1213–21.
[40] Simmons CP, et al. Dengue. N Engl J Med 2012;366(15):1423–32.
[41] Plennevaux E, et al. Impact of dengue vaccination on serological diagnosis: insights from phase III dengue vaccine efficacy trials. Clin Infect Dis 2018;66(8):1164–72.
[42] Jessie K, et al. Localization of dengue virus in naturally infected human tissues, by immunohistochemistry and in situ hybridization. J Infect Dis 2004;189(8):1411–8.
[43] Yamada K, et al. Virus isolation as one of the diagnostic methods for dengue virus infection. J Clin Virol 2002;24(3):203–9.

[44] Walker T, et al. Mosquito cell lines: history, isolation, availability and application to assess the threat of arboviral transmission in the United Kingdom. Parasites Vectors 2014;7. p. 382-382.
[45] Jarman RG, et al. Factors influencing dengue virus isolation by C6/36 cell culture and mosquito inoculation of nested PCR-positive clinical samples. Am J Trop Med Hyg 2011;84(2):218–23.
[46] Moi ML, et al. Dengue virus isolation in mosquito *Aedes albopictus* captured during an outbreak in Tokyo, 2014, by a method relying on antibody-dependent enhancement mechanism using FcgammaR-expressing BHK cells. Vector Borne Zoonotic Dis 2016;16(12):810–2.
[47] Medina F, et al. Dengue virus: isolation, propagation, quantification, and storage. Curr Protoc Microbiol 2012 [Chapter 15]: p. Unit 15D.2.
[48] Lebuhn M, et al. DNA and RNA extraction and quantitative real-time PCR-based assays for biogas biocenoses in an interlaboratory comparison. Bioengineering (Basel) 2016;3(1):7.
[49] Dos Santos HW, et al. A simple one-step real-time RT-PCR for diagnosis of dengue virus infection. J Med Virol 2008;80(8):1426–33.
[50] Kong YY, et al. Rapid detection, serotyping and quantitation of dengue viruses by TaqMan real-time one-step RT-PCR. J Virol Methods 2006;138(1–2):123–30.
[51] Wu SJ, et al. Detection of dengue viral RNA using a nucleic acid sequence-based amplification assay. J Clin Microbiol 2001;39(8):2794–8.
[52] Muñoz-Jordán JL, et al. Highly sensitive detection of dengue virus nucleic acid in samples from clinically ill patients. J Clin Microbiol 2009;47(4):927–31.
[53] van der Vliet GM, et al. Nucleic acid sequence-based amplification (NASBA) for the identification of mycobacteria. J Gen Microbiol 1993;139(10):2423–9.
[54] Notomi T, et al. Loop-mediated isothermal amplification of DNA. Nucleic Acids Res 2000;28(12). p. E63-E63.
[55] Wong YP, et al. Loop-mediated isothermal amplification (LAMP): a versatile technique for detection of micro-organisms. J Appl Microbiol 2018;124(3):626–43.
[56] Notomi T, et al. Loop-mediated isothermal amplification of DNA. Nucleic Acids Res 2000;28(12):E63.
[57] Hermann LL, et al. Evaluation of a dengue NS1 antigen detection assay sensitivity and specificity for the diagnosis of acute dengue virus infection. PLoS Neglected Tropical Diseases 2014;8(10). e3193-e3193.
[58] Anand AM, et al. Evaluation of NS1 antigen detection for early diagnosis of dengue in a tertiary hospital in Southern India. J Clin Diagn Res 2016;10(4):DC01–4.
[59] Chow L, Hsu ST. MAC-ELISA for the detection of IgM antibodies to dengue type I virus (rapid diagnosis of dengue type I virus infection). Zhonghua Min Guo Wei Sheng Wu Ji Mian Yi Xue Za Zhi 1989;22(4):278–85.
[60] Sathish N, et al. Comparison of IgM capture ELISA with a commercial rapid immunochromatographic card test & IgM microwell ELISA for the detection of antibodies to dengue viruses. Indian J Med Res 2002;115:31–6.
[61] Hunsperger EA, et al. Evaluation of commercially available anti-dengue virus immunoglobulin M tests. Emerg Infect Dis 2009;15(3):436–40.
[62] Vázquez S, et al. Diagnosis of dengue virus infection by the visual and simple AuBioDOT immunoglobulin M capture system. Clin Diagn Lab Immunol 2003;10(6):1074–7.
[63] Parikh F. Infections and thrombocytopenia. J Assoc Physicians India 2016;64(2):11–2.

[64] Cao XT, et al. Evaluation of the World Health Organization standard tourniquet test and a modified tourniquet test in the diagnosis of dengue infection in Viet Nam. Trop Med Int Health 2002;7(2):125–32.
[65] Potts JA, Rothman AL. Clinical and laboratory features that distinguish dengue from other febrile illnesses in endemic populations. Trop Med Int Health 2008;13(11):1328–40.
[66] Ralapanawa U, et al. Value of peripheral blood count for dengue severity prediction. BMC Res Notes 2018;11(1). 400-400.
[67] Wiwanitkit V, Manusvanich P. Can hematocrit and platelet determination on admission predict shock in hospitalized children with dengue hemorrhagic fever? A clinical observation from a small outbreak. Clin Appl Thromb Hemost 2004;10(1):65–7.
[68] Trung DT, et al. Liver involvement associated with dengue infection in adults in Vietnam. Am J Trop Med Hyg 2010;83(4):774–80.
[69] Neeraja M, et al. Unusual and rare manifestations of dengue during a dengue outbreak in a tertiary care hospital in South India. Arch Virol 2014;159(7):1567–73.
[70] Johnson AJ, et al. Duplex microsphere-based immunoassay for detection of anti-West Nile virus and anti-St. Louis encephalitis virus immunoglobulin m antibodies. Clin Diagn Lab Immunol 2005;12(5):566–74.
[71] Johnson AJ, et al. Validation of a microsphere-based immunoassay for detection of anti-West Nile virus and anti-St. Louis encephalitis virus immunoglobulin m antibodies. Clin Vaccine Immunol 2007;14(9):1084–93.
[72] Xia H, et al. A microsphere-based immunoassay for rapid and sensitive detection of bovine viral diarrhoea virus antibodies. J Virol Methods 2010;168(1–2):18–21.
[73] Lin Y-L, et al. A new microsphere-based immunoassay for measuring the activity of transcription factors. Biol Proced Online 2010;12(1):18–26.
[74] Vigneshvar S, et al. Recent advances in biosensor technology for potential applications – an overview. Front Bioeng Biotechnol 2016;4. p. 11-11.
[75] Salek-Maghsoudi A, et al. Recent advances in biosensor technology in assessment of early diabetes biomarkers. Biosens Bioelectron 2018;99:122–35.
[76] Mehrotra P. Biosensors and their applications – a review. J Oral Biol Craniofac Res 2016;6(2):153–9.
[77] Govindarajan R, et al. Microarray and its applications. J Pharm BioAllied Sci 2012;4(Suppl. 2):S310–2.
[78] Raghavachari N. Microarray technology: basic methodology and application in clinical research for biomarker discovery in vascular diseases. Methods Mol Biol 2013;1027:47–84.
[79] Díaz-Badillo A, et al. A DNA microarray-based assay to detect dual infection with two dengue virus serotypes. Sensors 2014;14(5):7580–601.
[80] Taraphdar D, et al. A comparative study of clinical features between monotypic and dual infection cases with chikungunya virus and dengue virus in West Bengal, India. Am J Trop Med Hyg 2012;86(4):720–3.
[81] Lee VJ, et al. Simple clinical and laboratory predictors of Chikungunya versus dengue infections in adults. PLoS Negl Trop Dis 2012;6(9):e1786.
[82] Waggoner JJ, et al. Viremia and clinical presentation in Nicaraguan patients infected with zika virus, chikungunya virus, and dengue virus. Clin Infect Dis 2016;63(12):1584–90.
[83] Naseer K, et al. Raman spectroscopy based differentiation of typhoid and dengue fever in infected human sera. Spectrochim Acta A Mol Biomol Spectrosc 2019;206:197–201.
[84] Jensenius M, et al. African tick bite fever. Lancet Infect Dis 2003;3(9):557–64.
[85] Tangpukdee N, et al. Malaria diagnosis: a brief review. Korean J Parasitol 2009;47(2):93–102.

Chapter 8

# Psychological and social aspects of Dengue virus illness virus infection

**Ihtesham Qureshi[1], Arbaab Qureshi[2]**
[1]*Resident physician, Department of Neurology, Texas Tech University of Health Sciences Center, Paul L. Foster School Of Medicine, El Paso, TX, United States;* [2]*Clinical research fellow, Department of Neurology, Texas Tech University of Health Sciences, El Paso, TX, United States*

*If you thought Dengue Virus Illness was about fever and falling platelet count, think again. There's a mental hazard associated with this affliction, one that is serious enough to make the government sit up and take notice*

——— *News Reporter highlighting the substantial increase in the number of dengue virus illness patients suffering from major psychiatric conditions in India after dengue virus illness outbreak [1].*

Dengue virus illness infection is a viral disease transmitted by Aedes mosquito and is endemic in tropical and subtropical regions of the world. Dengue virus illness infection is considered a global public health threat accounting for about 50% of world's population at a potential risk of having the infection [2].

Within the last couple of years, there have been collective reports of Dengue virus illness presenting with atypical features, predominantly psychiatric symptoms. Several studies have established that there is a strong linkage between the onset of Dengue virus illness and the appearance of manic symptoms [3].

There are several case reports highlighting the emergence of psychiatric symptoms of Dengue virus illness, which presents early onset of manic symptoms, observed during febrile and convalescence phase of Dengue virus illness. Initial manifestations include talkativeness, authoritative, and short-tempered conduct. In addition, patients may exhibit grandiosity, increased sexual behavior, and decreased need to sleep. To regulate these symptoms, mood stabilizers and antipsychotics are prescribed, which usually cause almost complete disappearance of symptoms within a short period of week to a month [4].

There is a strong association between physical and mental health. People living with mental issues have a higher chance of suffering from physical health issues, the same goes vice-versa, where people living with long lasting physical

**Dengue Virus Disease. https://doi.org/10.1016/B978-0-12-818270-3.00008-4**

health issues face widespread psychological conditions of which depression and anxiety are most commonly reported in comparison to general population. This is validated by a study performed by Jahnjee et al. [5] reported substantial psychiatric morbidity among patients affected with Dengue virus illness, even among them, majority of symptoms were observed in the early phase of Dengue virus illness which includes fear of death followed by anxiety. Of those patients recovered from Dengue virus illness, 35% exhibited persistent insect phobia and 7% reported death phobia at the end of sixth week [5].

Similar studies performed by Hashmi et al. and Gill et al. [6,7] also reported similar findings especially in the early phase of dengue virus illness where patients reported thanatophobia (fear of death) among nearly 90% of patients. Panic attacks in 15%–23% and approximately 80% had anxiety-related symptoms. As the physical health improved, the intensity and regularity of the psychiatric symptoms improved concurrently. Of those who recovered, about 50% developed insect phobia. Also, depressive symptoms were initially seen in 50%–60% of patients with 5%–20% of patients ending up with persistent symptoms. These psychiatric symptoms were most predominantly observed among women [6,7]. In a study performed by Gill et al. to assess the psychological effects of Dengue virus illness reported a finding that is worth mentioning that media, both print and electronic played a major role in the creation of psychological distress particularly fear of death. When questioned the real cause behind their extreme fear, they attributed this to the "breaking news" appearing on television constantly, along with scrolls and highlighting the deaths caused by Dengue virus illness in a dreadful way almost every day has created this fear among the people. This fear further multiplied in leaps and bounds when they found out that they have been affected by Dengue virus illness [7].

To understand this issue in a better way, there was an interesting study performed by Wong et al. on a focused group study in Malaysian citizens to understand their health benefits and practices related to Dengue virus illness prevention. The study reported that participants mainly fell into two different attitude themes in regards to Dengue virus illness: serious or highly deadly, and not a threat. The first category are participants that comprise about 33% of total participants, of which most knew of neighbors, friends, or colleagues who passed away from Dengue virus illness considered the infection as a dreadful and dangerous disease. These first category participants regarded Dengue virus illness as a disease that can kill people really fast as they heard from people who they know died in a short time soon after getting admitted to hospital. The rest of the participants (second category) that comprise about 66% regarded Dengue virus illness as not a hazard. These second category participants were mostly younger persons between 18 and 35 years old who believe that Dengue virus illness is not a dreadful disease and if diagnosed and treated early people can survive the disease.

Most participants consider Dengue virus illness a common and widely prevalent disease in most areas of Malaysia. A small portion of these participants believe that the chance of contracting the disease is next to negligible and those people who get affected from the disease are mainly from bad luck, ill fate, or chance. These participants further rationalize that only Aedes mosquito can transmit the infection and all other kinds of species are not capable for disease transmission. Furthermore, even among the Aedes mosquitoes, only those infected by dengue virus has the potential to cause Dengue virus illness. Few participants think their risk of attaining Dengue virus illness is extremely low as they are almost certain that the mosquitoes in their area were not Aedes species.

*The mosquitoes here are from the jungle, they are not dangerous. Every evening they come, we are used to it, mosquito bites are common, we turn on the fan every night.*

Malay female, 50 years old, housewife.

Additionally, there are a lot of participants, majority of them are young men, who has a perception that, if someone has a strong immune system, they cannot be affected by Dengue virus illness. Only those people with decreased immunity will be susceptible for contracting the Dengue virus illness fever.

*The place I stay … very likely …. because there is a lake, water stagnant. Many cases, I may get Dengue Virus Illness, but I think I am at low risk, because my body defense is strong and I am healthy.*

Indian male, 21 years old, college student.

One of the participants voiced that mosquitoes mostly favor young population over elderly, as the presence of thickened skin layer among elderly, is what will serve as a protection against mosquito bites.

*The mosquitoes prefer to bite small children than old people. Our skin thick and hard, cannot go through.*

Malay male, 69 years old [8].

## Social implications of dengue virus illness virus disease

The influence of the Dengue virus epidemic has profound and grave implications on the functionality of multiple organizations at various levels.

## Effect on health care infrastructure and management

*It was scary. Both public and private service centers were overflowing. The urgent care centers in private hospitals were also over capacity*

The response above clearly indicates the anxiety and panic the patients affected by the Dengue virus illness epidemic experienced during their visit to the health care facilities.

Lack of required space and equipment to accommodate and treat has forced the health care facilities to shift the patients or refer them to other hospitals further away from their homes, causing further inconvenience.

The environment created by the outbreak made the case notifications and follow-up processes more difficult.

> *To notify a certain pathology, you need to have time. Many times, the number of patients that you see is so extensive that you do not have the time to fill out a mandatory notification form for each patient. There is a huge demand and I think that it's because of it that the number of Dengue Virus Illness cases are not always notified properly*

The physicians and the hospital authorities reported a strong spike in the informal complaints and protests expressed by the patients and community members, which were related to procedural changes and longer waiting periods.

## Effect on households

Community leaders reported the dire consequences were much more apparent if the infected member of the household were women head as opposed to men.

> *The impact when a female head of household is diagnosed with Dengue Virus Illness is greater than a male head of household is diagnosed with Dengue Virus Illness because she is usually responsible for running the household.*

Overall, financial impact of dengue virus illness outbreaks on households was limited, but its effect was greatest for low-income families. Community members reported purchasing a range of mosquito control methods during an outbreak, such as citronella candles, insect swatters, special plants, sand to fill empty receptacles and insect repellants. It has been reported that the higher income households used insect repellant as the method of prevention, whereas nothing substantial was reported among the lower income households. Even though, majority of the health care facilities provided free treatment to the infected individuals, most of the patients tried self-medicating to alleviate the symptoms during the beginning of the outbreak. However, they were compelled to seek medical attention once the symptoms failed to subside or had gotten worse.

## Effect on schools

School authorities reported higher number of absenteeism during Dengue virus illness outbreaks. Increased absenteeism was recorded among students and teachers with the percentage ranging from 10% to 15%, respectively.

School administrators noted that schools lacked additional funding and necessary support for Dengue virus illness prevention and control measures.

No additional funding were received by the schools for Dengue virus illness prevention and control measures. To address the need for community sensitization during a Dengue virus illness outbreak, schools adapted new curriculums. Educational materials and lessons on Dengue virus illness prevention and control were introduced into the curriculum, where the focus was primarily on identification and removal of mosquito breeding sites and personal care. Several meetings were organized where parents were educated and given an insight about prevention of Dengue Virus Illness. They were also reassured that appropriate measures were being taken to control the spread of the infection.

## Effect on government and role of media

*A Dengue Virus Illness outbreak can lead to a political breakdown, despite the best efforts of leaders*

In response to the growing demand for care and treatment during an outbreak and the resulting community dissatisfaction with government dengue virus illness prevention and control efforts, governments intensified their efforts. In order to accommodate the increasing demands for care and treatment, and the deepening dissatisfaction among the community, the governments reinforced and intensified their efforts to prevent and control the Dengue virus illness. States of emergencies were declared during Dengue virus illness outbreaks and additional powers were given to the governments to take necessary actions aiming toward vector control activities and policies.

The political structures were significantly impacted due to the growing fear and anxiety of contracting the Dengue virus illness. Community discontent and apprehension for government's efforts in the prevention and control efforts heightened and, incited a high level of community disengagement with these efforts. The media played a prominent role in spreading the news and the reaction of the community in response to the outbreak.

The International media primarily focused on the consequences of the outbreak and the repercussions on the growing political tensions. The national and local media reported the government announcements, release of new epidemiological bulletins, and the direct consequences of the outbreak among the communities.

The leaders of the governments recognized the impact of the Dengue virus illness outbreaks and the serious implications this could cause which perhaps may lead to consequential political disintegration [9].

## Conclusion

Educating the population about how Dengue virus illness spreads is the key to reducing stress and anxiety. Dengue virus illness escalated to a social frenzy due to the neglect of the community leaders and government agencies to educate the public. The government and community leaders need to take responsibility and aid in the proper outflow of information to the public. It is therefore essential that leaders from the community and government postings participate in dissemination of information including the right/left wing activists, religious scholars, and government agencies. Without a combined effort to properly dispel the misinformation, the public will continue to live in a frenzy without seeking proper aid or treatment. Studies demonstrate that the sections of the public population that responded best to the Dengue virus illness outbreak and followed appropriate practices for dengue virus illness control were those that were well-organized, having active leaders, participated with government/ Nongovernmental Organization (NGO) in awareness campaign for Dengue virus illness eradication. These efforts include educating the public about the appropriate use of insecticides, sharing information regarding Dengue virus illness, and keeping mutual coordination with health department/ NGOs or other agencies in Dengue virus illness control [10].

Ideally, community participation is the strongest in countries that have stable political systems. Mobilization of the community needs to occur from the national level as well as the grassroots level to effectively prevent and control Dengue virus illness. One manner in which the local authorities can contribute is by eliminating breeding places of Dengue virus illness mosquitoes. This is the only cost-effective and sustainable way of ensuring control in any Dengue virus illness-affected country, especially those deficient in resources [11]. One such success story is the "**Thai National Dengue Virus Illness Prevention and Control Plan**" that was helpful in guiding the public protect and prevent against Dengue virus illness [12]. Successful community mobilization cannot occur without decentralization of resources and power. It also requires a high level of coordination between the agencies and the public. The first step should be to understand the daily problems of the community and then delegate proper prevention techniques. Without proper coordination, all efforts to control Dengue virus illness may be ineffective and costly. Studies demonstrate the need for proper planning and management in water supply, the drainage system and discarding broken items, are effective steps for successful Dengue virus illness control [13].

A similar success story was witnessed in Singapore by "DO THE MOZZIE"campaign launched on 28 April 2013 and since then, the annual Dengue virus illness prevention campaign calls for the community to actively check for and get rid go stagnant water in their homes by participating the 5-STEP MOZZIE WIPEOUT. The campaign is supported by local grassroots advisors and the community, with the mobilization of grass root leaders and the Dengue virus illness prevention volunteers

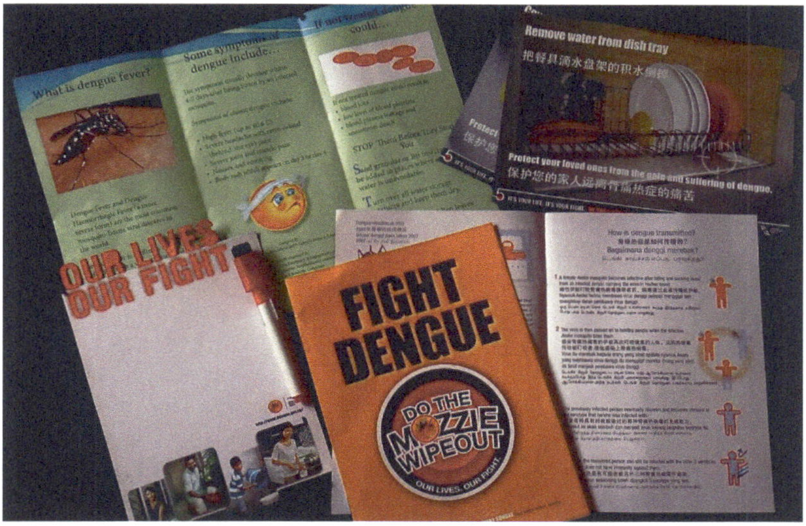


**FIGURE 8.1** Pamphlets distributed in 2013 by the new (National Environment agency) Singapore as part of the "do the Mozzie Wipeout" campaign.

(DPVs) to conduct house visits and organize Dengue virus illness outreach events (Fig. 8.1) [14].

The spread of Dengue virus illness can be controlled by the community through eliminating the Dengue virus vectors. The Dengue virus illness prevention campaign insists that the community keeps their living environment clean, removes standing water and burry discard broken items to eradicate the Dengue virus illness virus vectors [10]. These findings are similar to Spiegel et al. [15] who states that Dengue virus illness vectors can be eliminated through the hard work of the community people, leaders and government agencies. This study demonstrates the importance of the Dengue virus illness campaign and how being informed about Dengue virus illness fever play a role in maintaining control of the Dengue virus illness outbreak [15].

To deal with the issue in a better way and gain confidence of the local community members especially with regards to vector control, guidelines precisely intended for each environment should be developed in specific. The best practices must be swiftly executed at the commencement of the outbreak to achieve maximum effect [16]. Preparation should include novel approaches to educate community people with regards to Dengue virus illness, increasing the accessibility of various resources before outbreaks and sustaining essential resources and amenities in between the outbreaks. Essentially during the outbreaks, real emphasis should be placed on eliminating mosquito breeding

FIGURE 8.2 The community leaders in Hadhramaut, Yemen getting educated about mosquito breeding sites and the importance of preventing disease transmission during the investigation of dengue virus illness outbreak.

sites, which may require mobilizing and educating the local community members (Fig. 8.2).

The best way to manage and increase confidence of community members during an outbreak is to have precise guidelines toward vector control, suited for different types of environment, ready for rapid implementation [16]. Planning should include educating community members and leaders about Dengue virus illness, increasing the availability of Dengue virus illness resources before outbreaks and maintaining necessary resources and services between outbreaks.

During outbreaks, local sensitization campaigns must focus on community mobilization activities to eradicate mosquito breeding sites (Fig. 8.3).

Multiple factors need to fall in place to properly control a Dengue viral Illness outbreak. Adequate knowledge or awareness is a major factor in the successful implementation of approaches used to eliminate Dengue virus illness. However, knowledge of Dengue Virus Illness prevention is not enough. Policy makers and planners need to get involved and run sensitization campaigns on lifestyle changes the public must adopt to encourage preventative measures. Strong prevention campaigns that are implemented at the right time will help the public gain confidence in government responses to Dengue viral illness outbreaks and decrease the chances of political breakdown [9,17].

FIGURE 8.3 The residents of Yemen conduct a Dengue virus illness fever investigation by examining stagnant water for aides mosquito larvae which is responsible for transmitting the Dengue virus illness fever.

The World Health Organization (WHO) has therefore come up with new strategies, in order for countries to achieve zero mortality from dengue virus illness and for this they recommend, a country must.

Improve case management and diagnosis to prevent deaths from Dengue virus illness by the following:

- Improving early clinical case detection, especially for Dengue virus illness with warning signs and severe Dengue virus illness;
- Improving management of severe cases with appropriate interventions especially careful intravenous rehydration and a greater evidence base for interventions.

Improve capacities to facilitate a reduction in the burden of the disease:

- Improve health service organization, including access and triage in endemic countries to prevent dengue virus illness related deaths;
- Improve health service reorganization for managing outbreak situations;
- Build capacity and establish quality assurance in both private and public sector;
- Develop evidence-based informed training material, including Dengue viral illness courses;
- Prepare for the arrival of vaccines and their public health implications, bearing in mind the challenges of widespread implementation [18].

If the above-mentioned strategies and protocols are efficiently implemented by the organizations, health officials, governmental agencies and community people at large at various levels, the psychological stress and anxiety and social implications surrounding dengue virus illness virus disease can be contained to a larger extent.

## References

[1] Chandra N. Patients to get mandatory counselling as figures reveal almost eighty percent suffer from anxiety. 2013. http://www.dailymail.co.uk/indiahome/indianews/article-2419416/Dengue Virus Illnesss-hidden-toll-mental-health-Patients-mandatory-counselling-figures-reveal-EIGHTY-cent-suffer-anxiety.html.

[2] Gunathilaka N, Chandradasa M, Champika L, Siriwardana S, Wijesooriya L. Delayed anxiety and depressive morbidity among Dengue Virus Illness patients in a multi-ethnic urban setting: first report from Sri Lanka. Int J Ment Health Syst 2018;12:20. https://doi.org/10.1186/s13033-018-0202-6.

[3] Jhanjee A, Bhatia MS, Srivastava S. Mania in dengue virus illness fever. Ind Psychiatry J 2011;20(1):56–7. https://doi.org/10.4103/0972-6748.98418.

[4] Bhatia MS, Saha R. Neuropsychiatric manifestations in dengue virus illness fever. Med J DY Patil Univ 2017;10:204–6. http://www.mjdrdypu.org/text.asp?2017/10/2/204/202111.

[5] Jhanjee A, Bhatia MS, Srivastava S, Rathi A. A study of psychiatric symptomatology in dengue virus illness patients. Delhi Psychiatry J 2013;16:21–3.

[6] Hashmi AM, Butt Z, Idrees Z, Niazi M, Yousaf Z, Haider SF, et al. Anxiety and depression symptoms in patients with dengue virus illness fever and their correlation with symptoms severity. Int J Psychiatry Med 2012;44:199–210.

[7] Gill KU, Ahmad W, Irfan M. A clinical study to see the psychological effects of dengue virus illness fever. Pak J Med Health Sci 2011;5:101–4.

[8] Wong LP, AbuBakar S. Health beliefs and practices related to dengue virus illness fever: a focus group study. PLoS Neglected Trop Dis 2013;7(7):e2310. https://doi.org/10.1371/journal.pntd.0002310.

[9] Ladner J, Rodrigues M, Davis B, Besson M-H, Audureau E, Saba J. Societal impact of Dengue Virus Illness outbreaks: stakeholder perceptions and related implications. A qualitative study in Brazil, 2015. PLoS Neglected Trop Dis 2017;11(3):e0005366. https://doi.org/10.1371/journal.pntd.0005366.

[10] Zahir A, Ullah A, Shah M, Mussawar A. Community participation, dengue virus illness fever prevention and control practices in swat, Pakistan. Int J Mch AIDS 2016;5(1):39–45.

[11] Khun S, Manderson L. Community participation and social engagement in the prevention and control of Dengue Virus Illness fever in rural Cambodia. Dengue Virus Illness Bulletin 2008;32:145–55.

[12] Kantachuvessiri A. Dengue Virus Illness hemorrhagic fever in Thai society. Southeast Asian J Trop Med Public Health 2002;33:56–62. Bangkok Thailand: faculty of public health, Mahidol university.

[13] Claro LB, Kawa H, Cavalini LT, Rosa MLG. Community participation in dengue virus illness control in Brazil. Dengue Virus Illness Bulletin 2006;30:214–22.

[14] National Environment Agency. 2018. https://www.nea.gov.sg/programmes-grants/campaigns/do-the-mozzie-wipeout-campaign.

[15] Spiegel JS, Bennett HL, Hayden MH, Kittayapong P, et al. Barriers and bridges to prevention and control of dengue virus illness: the need for a social—ecological approach. EcoHealth 2012;2:273—90.

[16] McNaughton D, Duong TT. Designing a community engagement framework for a new dengue virus illness control method: a case study from Central Vietnam. PloS Negl Dis 2014;8:e2794.

[17] Runge-Ranzinger S, McCall PJ, Kroeger A, Horstick O. Dengue virus illness disease surveillance: an updated systematic literature review. Trop Med Int Health 2014;19:1116—60.

[18] Global strategy for dengue virus illness prevention and control. 2012—2020. World Health Organization (WHO), http://apps.who.int/iris/bitstream/handle/10665/75303/9789241504034_eng.pdf;jsessionid=D1B8C9859E520412654E9E8D0EC415CC?sequence=1.

Chapter 9

# Economic and political aspects of Dengue virus disease

**Mushtaq H. Qureshi**
*Texas Tech University Health Sciences Center El Paso, Neurology, Texas Tech University, El Paso, TX, United States; Zeenat Qureshi Stroke Institute, St Cloud, MN, United States*

## Disease burden

The disease and economic burdens of Dengue virus infection are considerable. It is essential to quantify these burdens by the policy-makers in order to allocate resources, to select prevention and control strategies, to set priorities, and to most importantly to evaluate the cost effectiveness of interventions [1]. Disease burden can be defined as the number of infections, years of life lost to premature mortality, years that are lived with disability, and/or life years that are disability-adjusted. The cost of disease burden can be conceptualized based on the disease effect on the society, which is associated with diagnosis, treatment, outcome, and disease prevention. The disease burden of Dengue virus infection has been estimated using different data, analytical methods, and geographical coverage. According to World Health Organization, there are 50–100 million annual cases of Dengue virus infection [2], whereas according to a research in 2013 by University of Oxford and Wellcome trust, it is estimated to be 390 million Dengue virus infections per year (Fig. 9.1). The estimated burden of Dengue virus infection based on geostatistical models and 2010 population data is to be 390 million (95% credible interval [CrI] 284–528) infections per year, of which 96 million (95% CrI 67–136) are those which are causing clinical symptoms [3]. Approximately, 70% of the disease burden is bourne by Asia in its large, highly populated areas (approximately 67 million infections annually). When compared with the official reported figures by the countries, these numbers are much larger. The Global Burden of Disease Study 2013 estimates a temporal trend in the burden of Dengue virus infection, with the number of apparent cases more than doubling each decade between 1990 and 2013 [4]. If we look at the children in particular, it is not only a higher burden of symptomatic infections, but there is a significant numbers of these infections which resulted in hospital admission

**Dengue Virus Disease. https://doi.org/10.1016/B978-0-12-818270-3.00009-6**

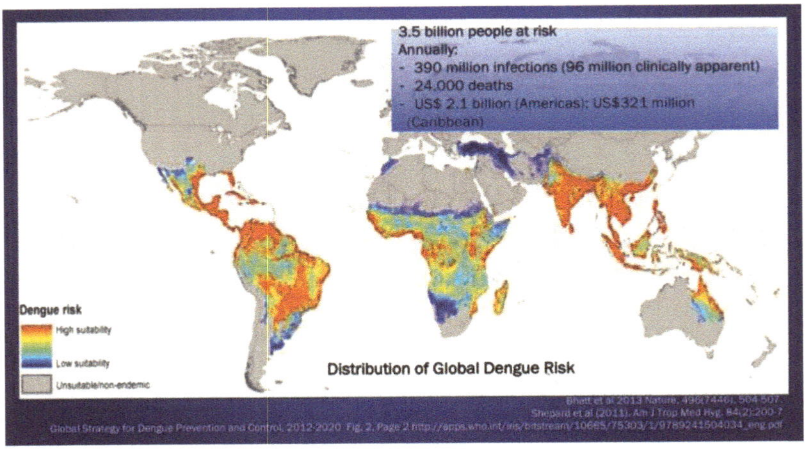


FIGURE 9.1 Distribution of global Dengue virus infection risk.

(ranging from 4.9% to 45.5% in 10 Asian and Latin American countries) [5]. Looking at the deaths, however, the estimates seem low and do not go with the increasing trend, which is observed in the apparent cases [4,6]. The Global Burden of Disease Study 2013 estimates that Dengue virus infection is responsible for 1.14 million disability-adjusted life-years in 2013, which is a clear 61% increase from 1990 [4,7]. Many countries, where Dengue virus infection is endemic, have insufficient data about deaths [4]. Data analyzed from a surveillance system from Puerto Rico recorded the highest Dengue virus infection mortality rate ever detected (1.05 per 100 000 people), and the rates were higher in adults with comorbidities. (1.66 per 100,000 people aged 65 years or older) [8]. These numbers, however, are still an underestimate. So far, only two studies have provided the global estimations of the disease burden [9,10]. Combining the disease burden estimates provided by Bhatt and colleagues(3) and the cost associated with Dengue virus infection treatment and loss of productivity (as provided by World Health Organization), the global estimated cost was estimated to be approximately US$39.3 billion (about $414 per symptomatic case) for 2011 [9]. Another study performed by Shepard and Colleagues (9) reviewed and combined data from different sources (including the Global Burden of Disease Study 2013, household data, expert panel surveys, and empirical cost data) in a modeling exercise that was able to produce the first worldwide estimation regarding the economic burden of Dengue virus infection, which was comparable across regions and countries [9]. This global health burden was estimated to be 58.4 million symptomatic cases, resulting in an estimated global cost of $8.9 billion (95% uncertainty interval 3.7−19.7) [9] which is higher than various other infectious diseases (Fig. 9.2).

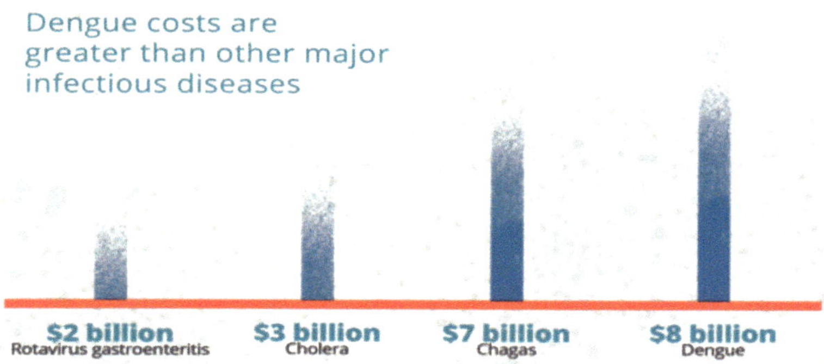


**FIGURE 9.2** Cost associated with Dengue virus infection compared with major infectious diseases.

> *Defining Dengue in monetary terms means the disease can be compared with other economic problems. Public health systems can then leverage that information to secure resources from their Ministry of Finance—and possibly the donor community—to control the disease.*
>
> Professor Donald Sheperd, PhD. (lead author and economist).

## Challenges in measuring the burden of disease

The challenges of measuring the burden of Dengue viral illness often result in uncertain estimates and hamper cross-country comparisons. Common challenges can be grouped into the following categories: finding out true number of cases, estimating the heterogeneous costs, quantifying the cyclical variations, and assessment of the burden. Usually dengue virus infection cases and death numbers that are reported by the endemic countries is an underestimate. The reason being, private health practices only provide limited information [11]; inability to detect Dengue virus infection in symptomatic patients seeking care due to surveillance gaps; when patients does not seek health care; financial restraints leading to restricted access to primary health care and no access to sensitive and specific diagnostic tests; in the health-care world when there is failure make a diagnosis of dengue virus infection [11]; lack of information technology and reliance on paper records; inadequate reporting of Dengue virus infection to the national authorities. Effect of Dengue virus infection on individual's productivity is also worth considering. For e.g., number of days lost at school or work is routinely not collected. Even more challenging is estimating the cost of reduced performance at work due to fatigue and other short-term and longer-term consequences of Dengue virus infection [12]. Quantifying cost during the outbreak can be very challenging for various reasons: during emergency periods when supplemental financial resources are disbursed, without detailed allocation to specific activities, it will make it difficult to categorize the different cost component and also the additional cost during out breaks is not easily quantifiable through routinely acquired data. According to a review [13], 50% or less studies estimating the economic burden of Dengue virus disease, based their reports on the basis of the costs of outbreaks.

## Impact on low-income and middle-income tropical countries

The economic burden of individual country or region often vary in several aspects: number of years of data used, types of costs, decisions on how to extrapolate the data, choice of cost analysis perspective, monetary reference, and geographical coverage. Estimates of the economic burden demonstrate that low-income and middle-income tropical countries are the ones, which are the most affected by Dengue virus infection. A major fraction (50%–60%) of the estimated economic costs of Dengue virus infection is related to loss of productivity [14,15], as well as vector control (40%–72% of the estimated cost) [1,16,17]. Vector control represents a significant portion of Dengue virus infection–related cost. The costs associated with Dengue virus infection in south Asian countries are comparatively higher when compared to other conditions such as, Japanese encephalitis, hepatitis B infections, upper respiratory infections, etc. [14] (Fig. 9.2) Also, when looking at the daily cost as a result of Dengue virus infection in India, the cost is twice the cost per day than of tuberculosis case [18]. When looking at the Americas, the economic burden of Dengue virus infection exceeded that of other viral diseases, e.g., human papillomavirus [15]. Assessment of the economic cost of Dengue virus infection in Puerto Rico (14) showed that households incurred 48% and employers 7% of the total Dengue virus infection cost, whereas the government and insurers bore 24% and 22% of the cost, respectively. Households incurred 90% of the cost that was associated with fatal cases, 21% for hospital admission for a child and 37% of the costs for hospital admission for an adult, and 51% and 63% of the costs for ambulatory child and adult cases, respectively. Also, when estimates were drawn for the direct medical costs in India [18], a country bearing one third of global disease burden of Dengue virus infection [3], suggest that private sources, mostly households, bore 80% of the cost. However, the burden could vary by country, depending on the finance structure of the health-care system. Uncertainties, associated with the estimation of the burden of Dengue virus infection reflect problems with availability and quality of data, and also raise important points: 1. Undoubtedly, the burden is large and growing; 2. Uncertainties associated with estimation of disease burden (i.e., infections and deaths) carry over into the estimation of the economic burden of disease; and 3. The greater the uncertainty in estimates, the more serious are the challenges that policy-makers face in priorities setting, resources allocation, and interventions planning.

## A comparison with Ebola virus disease

According to the World Bank, budget deficits of the affected countries continue to increase by amount equal to 1.8% of gross domestic product in countries including Sierra Leone and Guinea and 4.7% in Liberia (Fig. 9.3). With the continuous virus surge in the three worst affected countries with span to neighboring countries, the 2-year regional financial impact by 2015 reached to be $32.6 billion, causing a potentially catastrophic blow to already fragile states. On the other hand, when comparing the Dengue virus infection disease burden estimates provided by Bhatt and colleagues(3) and the cost associated with its treatment and loss of productivity (as provided by World Health Organization), the global estimated cost was estimated to be approximately US$39.3 billion (about $414 per symptomatic case) for 2011.

## A comparison with Zika virus disease

Zika virus disease which is primarily spread by *Aedes aegypti* mosquito was declared a public health emergency of international concern in February of 2016 [19]. Although no longer considered a public health emergency of international concern, Zika virus disease is still considered as a health issue which has the potential to hit the most vulnerable communities the hardest. Several factors had played a role in the economic burden posed by this disease. According to the statement from World Bank Group, "Initial estimates of the short-term economic impact of the Zika virus epidemic for 2016 in the Latin American and the Caribbean region (LCR) are a total of US$3.5 billion, or 0.06% of GDP" (Table 9.1). What they used were a few factors to calculate

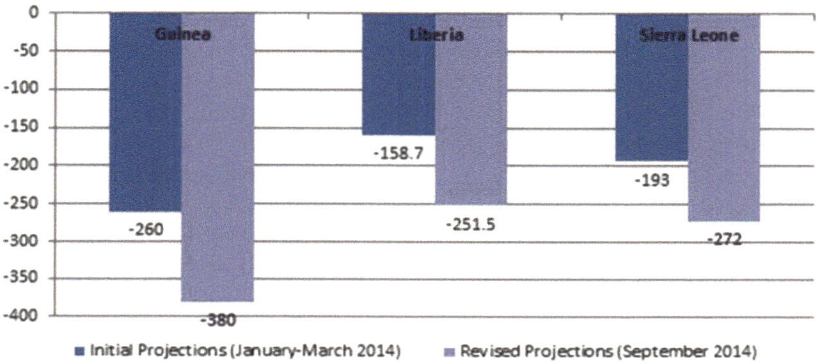


FIGURE 9.3 Estimated short-term impact on the overall fiscal balance (2014), in $US millions of ebola virus disease.

**TABLE 9.1 World Health Organization estimates of the short-term economic impact of the Zika virus epidemic for 2016 in the Latin American and the Caribbean region.**

| | |
|---|---|
| Mexico | $744 million |
| Cuba | $664 million |
| Dominican Republic | $318 million |
| Brazil | $310 million |
| Argentina | $229 million |
| Jamaica | $112 million |
| Belize | $21 million |
| Other | $1.08 billion |
| Total | $3.48 billion |

**Source:** World Bank.

these numbers. These factors included behaviors to avoid transmission, its effect on economic staples like tourism, loss of worker productivity, public perceptions of risk from Zika virus disease which included media attentions and the urgent need to take the action against the virus's spread.

## Dengue virus infection vaccine, a promising tool

A tool that has shown promise against Dengue virus infection is the use of a vaccine. Even if a vaccine is partially effective, it would still affect the burden of the disease, will decrease the number of new infections and therefore decreasing the costs associated with illness from the perspective of health providers, individuals or households, and society. A vaccine with effective control would be an excellent addition to the limited number of tools which are currently available to reverse the growing burden of Dengue virus infection [20]. Several countries (e.g., Brazil, Mexico, the Philippines, Indonesia, Costa Rica, Paraguay, and El Salvador) have approved the use of a live recombinant tetravalent Dengue virus infection vaccine that is administered in three doses in individuals between 9 and 45 years of age. Several phase 3 trials in Asia and Latin America have shown reductions in numbers of severe Dengue virus infection cases as well as the need for hospitalization among certain groups [21—23], but if we consider the protective efficacy, it is varied by serotype and presence of antibodies from previous infection at the time of vaccination. These trials have also shown possible disease enhancement in young children [23].However research continues with other vaccine candidates [24—27].

## References

[1] Carrasco LR, Lee LK, Lee VJ, Ooi EE, Shepard DS, Thein TL, et al. Economic impact of dengue illness and the cost-effectiveness of future vaccination programs in Singapore. PLoS Neglected Trop Dis 2011;5(12):e1426.

[2] Dengue: guidelines for diagnosis, treatment, prevention and control. New Edition. Geneva: WHO Guidelines Approved by the Guidelines Review Committee; 2009.

[3] Bhatt S, Gething PW, Brady OJ, Messina JP, Farlow AW, Moyes CL, et al. The global distribution and burden of dengue. Nature 2013;496(7446):504−7.

[4] Stanaway JD, Shepard DS, Undurraga EA, Halasa YA, Coffeng LE, Brady OJ, et al. The global burden of dengue: an analysis from the Global Burden of Disease Study 2013. Lancet Infect Dis 2016;16(6):712−23.

[5] L'Azou M, Moureau A, Sarti E, Nealon J, Zambrano B, Wartel TA, et al. Symptomatic dengue in children in 10 Asian and Latin American countries. N Engl J Med 2016;374(12):1155−66.

[6] Wilder-Smith A, Byass P. The elusive global burden of dengue. Lancet Infect Dis 2016;16(6):629−31.

[7] Hotez PJ, Alvarado M, Basanez MG, Bolliger I, Bourne R, Boussinesq M, et al. The global burden of disease study 2010: interpretation and implications for the neglected tropical diseases. PLoS Neglected Trop Dis 2014;8(7):e2865.

[8] Tomashek KM, Rivera A, Torres-Velasquez B, Hunsperger EA, Munoz-Jordan JL, Sharp TM, et al. Enhanced surveillance for fatal dengue-like Acute Febrile illness in Puerto Rico, 2010−2012. PLoS Neglected Trop Dis 2016;10(10):e0005025.

[9] Russo M, Bevilacqua P, Netti PA, Torino E. A Microfluidic platform to design crosslinked hyaluronic Acid nanoparticles (cHANPs) for enhanced MRI. Sci Rep 2016;6:37906.

[10] Selck FW, Adalja AA, Boddie CR. An estimate of the global health care and lost productivity costs of dengue. Vector Borne Zoonotic Dis 2014;14(11):824−6.

[11] Shepard DS, Undurraga EA, Betancourt-Cravioto M, Guzman MG, Halstead SB, Harris E, et al. Approaches to refining estimates of global burden and economics of dengue. PLoS Neglected Trop Dis 2014;8(11):e3306.

[12] Barnighausen T, Bloom DE, Cafiero ET, O'Brien JC. Valuing the broader benefits of dengue vaccination, with a preliminary application to Brazil. Semin Immunol 2013;25(2):104−13.

[13] Constenla D, Garcia C, Lefcourt N. Assessing the economics of dengue: results from a systematic review of the literature and expert survey. Pharmacoeconomics. 2015;33(11): 1107−35.

[14] Shepard DS, Undurraga EA, Halasa YA. Economic and disease burden of dengue in Southeast Asia. PLoS Neglected Trop Dis 2013;7(2):e2055.

[15] Shepard DS, Coudeville L, Halasa YA, Zambrano B, Dayan GH. Economic impact of dengue illness in the Americas. Am J Trop Med Hyg 2011;84(2):200−7.

[16] Undurraga EA, Betancourt-Cravioto M, Ramos-Castaneda J, Martinez-Vega R, Mendez-Galvan J, Gubler DJ, et al. Economic and disease burden of dengue in Mexico. PLoS Neglected Trop Dis 2015;9(3):e0003547.

[17] Castaneda-Orjuela C, Diaz H, Alvis-Guzman N, Olarte A, Rodriguez H, Camargo G, et al. Burden of disease and economic impact of dengue and severe dengue in Colombia, 2011. Value Health Reg Issues 2012;1(2):123−8.

[18] Shepard DS, Halasa YA, Tyagi BK, Adhish SV, Nandan D, Karthiga KS, et al. Economic and disease burden of dengue illness in India. Am J Trop Med Hyg 2014;91(6):1235−42.

[19] Organization WH. Fifth meeting of the Emergency Committee under the International Health Regulations (2005) regarding microcephaly, other neurological disorders and Zika virus. 2005.

[20] Rodriguez-Barraquer I, Mier-y-Teran-Romero L, Schwartz IB, Burke DS, Cummings DA. Potential opportunities and perils of imperfect dengue vaccines. Vaccine 2014;32(4): 514–20.

[21] Hadinegoro SR, Arredondo-Garcia JL, Capeding MR, Deseda C, Chotpitayasunondh T, Dietze R, et al. Efficacy and long-term safety of a dengue vaccine in regions of endemic disease. N Engl J Med 2015;373(13):1195–206.

[22] Villar L, Dayan GH, Arredondo-Garcia JL, Rivera DM, Cunha R, Deseda C, et al. Efficacy of a tetravalent dengue vaccine in children in Latin America. N Engl J Med 2015;372(2):113–23.

[23] Capeding MR, Tran NH, Hadinegoro SR, Ismail HI, Chotpitayasunondh T, Chua MN, et al. Clinical efficacy and safety of a novel tetravalent dengue vaccine in healthy children in Asia: a phase 3, randomised, observer-masked, placebo-controlled trial. Lancet 2014; 384(9951):1358–65.

[24] Durbin AP, Kirkpatrick BD, Pierce KK, Elwood D, Larsson CJ, Lindow JC, et al. A single dose of any of four different live attenuated tetravalent dengue vaccines is safe and immunogenic in flavivirus-naive adults: a randomized, double-blind clinical trial. J Infect Dis 2013;207(6):957–65.

[25] Osorio JE, Huang CY, Kinney RM, Stinchcomb DT. Development of DENVax: a chimeric dengue-2 PDK-53-based tetravalent vaccine for protection against dengue fever. Vaccine 2011;29(42):7251–60.

[26] Coller BA, Clements DE, Bett AJ, Sagar SL, Ter Meulen JH. The development of recombinant subunit envelope-based vaccines to protect against dengue virus induced disease. Vaccine 2011;29(42):7267–75.

[27] Beckett CG, Tjaden J, Burgess T, Danko JR, Tamminga C, Simmons M, et al. Evaluation of a prototype dengue-1 DNA vaccine in a Phase 1 clinical trial. Vaccine 2011;29(5):960–8.

Chapter 10

# Treatment and therapeutic agents and vaccines

Sargun Singh Walia[1,2], Ngan Nguyen[3], Mohammad F. Ishfaq[4,5]
[1]*Clinical Research Fellow, Zeenat Qureshi Stroke Institute, St. Cloud, MN, United States;* [2]*Department of Neurology, University of Missouri, Columbia, MO, United States;* [3]*Department of Internal Medicine, University of Tennessee Health Science Center, Memphis, Tennessee, United States;* [4]*Resident physician, University of Tennessee Health Science center, Memphis, Tennessee, United States;* [5]*Zeenat Qureshi Stroke Institute, St. Cloud, MN, United States*

## Management of Dengue viral illness

The mainstay of management for Dengue viral illness is supportive care. There is not antiviral treatment directed toward Dengue viral illness. As the main pathogenesis of Dengue viral illness is fever, plasma leakage and shock mainstay therapy is directed toward managing these.

### Overall assessment

As mentioned in any treatment guidelines published by World Health Organization (WHO; 2009) and World Health Organization South East Asia Regional Office, the first step in management involves taking a history from the patient about the symptoms along with past medical and family history.

A good history should include the onset of fever, mucosal bleeding or any internal organ bleeding, any urinary symptoms, change in mental status, seizure, dizziness and recent travel history including travel to or living in Dengue viral illness-endemic areas. Assessment of warning signs, coexisting conditions and poor social circumstances is important in determining the level of care (Table 10.1).

Second part of the initial assessment is physical examination. A thorough physical examination includes assessment of volume status, hemodynamic status, mental state, examination for rash and bleeding manifestation, checking for abdominal tenderness, hepatomegaly or ascites, any signs of respiratory distress including tachypnea, kussmaul breathing sign or pleural effusion on imaging. A mental assessment should also be performed along with physical examination (Table 10.2).

Dengue Virus Disease. https://doi.org/10.1016/B978-0-12-818270-3.00010-2

**TABLE 10.1 List of Warning Signs, Coexisting conditions and Poor Social Circumstances.**

| Warning signs | Coexisting conditions | Poor social circumstances |
|---|---|---|
| • Abdominal pain<br>• Persistent vomiting<br>• Clinical fluid accumulation<br>• Mucosal bleed<br>• Lethargy, Restlessness<br>• Liver enlargement >2 cm<br>• Laboratory showing elevated hematocrit with rapid declined platelet count. | • Infancy<br>• Pregnancy<br>• Obesity<br>• Diabetes mellitus<br>• Hypertension. | • Living alone<br>• Living far away from hospital. |

### *Laboratory test*

A complete blood count and comprehensive metabolic panel should be performed at the first visit to assess for any signs of leukopenia and thrombocytopenia as these make the diagnosis of Dengue viral illness highly likely. A rapid decline in platelet count with an increase in hematocrit above baseline indicates plasma leakage and progress to critical phase of the disease. In addition, patient with Dengue viral illness may develop electrolyte abnormalities, mildly elevated transaminitis and elevated serum creatinine level.

After a thorough examination and laboratory workup, clinicians should determine whether Dengue viral illness is the most likely disease, and phase of the disease (febrile, critical and recovery phase).

**TABLE 10.2 Important items of history and physical examination of Dengue viral illness patients.**

| History | Physical examination |
|---|---|
| • Fever onset<br>• Oral fluid intake<br>• Diarrhea<br>• Urine output<br>• Mental state change/seizure/dizziness<br>• Family history<br>• Travel history<br>• Sexual history<br>• Drug use | • Mental state assessment<br>• Hydration state assessment<br>• Hemodynamic status assessment<br>• Look for rash and bleeding manifestations.<br>• Tourniquet test<br>• Respiratory: Look for tachypnoea/pleural effusion<br>• Gastrointestinal: Abdominal tenderness/hepatomegaly/ascites |

The World Health Organization handbook for clinical management of Dengue 2012 has divided patients into three groups to help aid in the disease management.

The groups are:

1. Group A (outpatient management)—Dengue viral illness without warning signs
2. Group B (inpatient management)—Dengue viral illness with warning signs
3. Group C (emergency management)—severe Dengue viral illness

## Group A (outpatient management)

Patients who do not present with any warning signs or coexisting conditions and able to tolerate adequate volume of oral fluid and pass urine every 6 h can be managed in the outpatient setting [1]. Serial cell count using complete blood count should be followed. An increase in hematocrit and a decline in platelet count indicate presence of plasma leakage and increased risk of bleeding complication. Patients should be monitored closely for any warning signs as they may decline rapidly in the critical phase. Patients who present with symptoms of Dengue viral illness for more than 3 days should be monitored on a daily basis for disease progression. Evaluations are done based on reduction in white blood cells and platelet counts along with an increase in hematocrit levels. Patients with a stable hematocrit level can be managed as outpatient.

Adequate oral fluid intake is the mainstay of management in these patients and can decrease the rate of hospitalization. Oral fluids in the form of coconut water, rice water, barley water, oral rehydration solutions (ORS), fruit juices, and soup are advised. Carbonated drinks should be avoided as they are rich in sugar content and can exacerbate physiological stress hyperglycemia present in Dengue viral illness patients. Oral fluids should be taken until the urinary frequency increase to 4 to 6 times per day. A record of fluid intake and urine output is helpful. Bed rest is advised to minimize any trauma or bleeding complications.

High fever should be controlled with the use of paracetamol. Maximum dose in children is 10 mg/kg/dose, with an upper limit of 3–4 times in a 24 h period. Maximum dose of paracetamol in adults is 3 grams/day in a 24 h period. If the fever is still not controlled then sponging can be done with tepid water. A randomized, double-blind, placebo-controlled trial was conducted to test the use of paracetamol in children suffering from fever (>38°C per rectum) less than 4 days. It was found that children taking paracetamol demonstrated improvement in activity and comfort [2]. Aspirin and nonsteroidal anti-inflammatory agents are not advised as they can lead to gastritis and bleeding complications.

Special care should be taken to look for development of any signs of volume depletion, dry mouth, cold extremities or severe abdominal pain, persistent vomiting, mucosal or internal organ bleeding such as difficult breathing, decreased urination frequency, black stools, or coffee ground vomiting warrant prompt clinical evaluation. Patient should be instructed to return to hospital immediately if these warning signs develop (Table 10.3).

## Group B (inpatient management)

These are the patients that present with warning signs or have preexisting conditions or social circumstances that can make management of Dengue viral illness a complicated as an outpatient. The most important key to management in these patients is fluid resuscitation to prevent progression to a state of hemodynamic shock.

Appropriate use of intravenous fluids can be vital in the management of these patients. The initial step is to get a baseline hematocrit level to use as reference to manage the intravenous fluid therapy. Intravenous fluid therapy is started in the form of 0.9% normal saline or lactated Ringer. Therapy is started at an initial rate of 5−7 mL/kg/h for 1−2 h, then decreased rate of 3−5 mL/kg/h for 2−4 h, with further decrease to 2−3 mL/kg/h or less based on clinical response of the patient.

**TABLE 10.3** Dengue viral illness Group A management algorithm.

| | Dengue viral illness without warning signs |
|---|---|
| **Classification** | **Group A (outpatient care)** |
| **Criteria** | Patients without warning signs or<br>• Patients able to take oral fluids<br>• Patients able to urinate at least once every 6 h |
| **Lab tests** | • Full blood count<br>• Hematocrit |
| **Treatment** | • Bed rest<br>• Fluid intake<br>• Paracetamol<br>• Stable hematocrit patients can be sent home |
| **Monitoring** | • Daily disease progression<br>• Reduction of white blood cells<br>• Defervescence<br>• Warning signs<br>• Immediate return to the hospital if warning signs develop |

Adapted from World Health Organization: Handbook for Clinical Management of Dengue (2012).

After fluid replacement, the hematocrit should be repeated. If there is no or minimal increase then the fluids can be maintained at a reduced rate of 2–3 mL/kg/h for another 2–4 h. But if the vitals are not stable or the hematocrit is increasing then the rate of fluid should be increased to 5–10 mL/kg/hour. The clinical status should be reassessed and hematocrit levels reviewed again to change the fluid rate accordingly.

The minimum intravenous fluids required to maintain perfusion and urine output of 0.5 mL/kg/hour should be given to all patients. Intravenous fluids are usually not needed after 24–48 h. As the urine output and oral fluid intake improves the fluids should be reduced at a gradual rate.

Health care providers should monitor these patients suffering with Dengue viral illness with warning signs. Fluid balance charts should be maintained in detail. Peripheral perfusion and vital signs should be monitored every 1–4 h until the patient is stable and out of the critical phase. Urine output should be monitored every 4–6 h. Hematocrit levels should be recorded before starting fluid replacement therapy and also after administering fluids. Then it should be repeated every 6–12 h. Renal profile, liver profile, coagulation studies, and blood glucose should be recorded as indicated. There is no clinical advantage of colloid over crystalloid [3].

In patients of Dengue viral illness with coexisting conditions, in the absence of warning symptoms, the treatment plan is different. The patients are advised to take oral fluids but if not tolerated then intravenous fluids are started in the form of 0.9% saline or Ringer lactate with or without glucose. The rate of fluid is decided based on the ideal body weight (Table 10.4).

For adults with ideal body weight (IBW) $>$ 50 kg, 1.5–2 mL/kg can be used as quick calculation for maintenance of fluid per hour.

For adults with ideal body weight (IBW) $<$ 50 kg, 2–3 mL/kg can be used as quick calculation for maintenance of fluid per hour.

Healthcare providers should monitor these patients for temperature pattern, fluid intake and output volumes, volume and frequency of urine output, white blood cell count, platelet count, and hematocrit. Renal profile, liver profile, coagulation studies, and blood glucose should be recorded as indicated (Table 10.5).

## Group C (emergency care)

Patients with severe Dengue viral illness can be categorized based on the following:

- Severe Dengue viral illness leading to shock. This is characterized by circulatory collapse due to an increased systemic vascular permeability and severe plasma leakage.
- Respiratory distress due to fluid accumulation caused by severe plasma leakage.

**TABLE 10.4 Guide for maintenance intravenous fluid infusion.**

| Calculator for maintenance of intravenous fluid infusion | |
|---|---|
| Normal maintenance fluid per hour based on the Holliday Segar formula | |
| • 4 mL/kg/h for first 10 kg body weight<br>• + 2 mL/kg/h for next 10 kg body weight<br>• + 1 mL/kg/h for subsequent kg body weight | |
| In overweight/obese patients maintenance fluid is based on ideal body weight (IBW), using the following formula: | |
| Female | 45.5 kg + 0.91(height−152.4) cm |
| Male: | 50.0 kg + 0.91(height−152.4) cm |

Adapted from World Health Organization: Handbook for Clinical Management of Dengue (2012).

- Severe hemorrhagic symptoms
- Severe organ impairment

Patients presenting with these symptoms should be immediately admitted in a medical facility capable of blood transfusion. The mainstay in management is the use of fluid replacement. The preferred intravenous fluid therapy is crystalloid solution. Crystalloid solution used should be isotonic. Plasma leakage should be replaced immediately and rapidly with enough volume of crystalloid solution so that effective circulation is maintained. Colloid solution is the preferred fluid replacement therapy in cases with hypotensive shock. Hematocrit levels are assessed before and after the fluid resuscitation.

Fluid replacement therapy is continued for at least 24–48 h to maintain effective circulation. Blood group and cross match should be in all patients in shock due to Dengue viral illness. In patients in whom severe bleeding is occurring or the patients with unexplained hypotension and suspicion of severe hemorrhage prompt treatment with blood transfusion is recommended.

As a trial, boluses of 10–20 mL/kg fluid are administered for a short duration of time under vigilant supervision to look for development of pulmonary edema. It is important to keep these boluses of fluid free of glucose. In patients suffering from severe shock due to Dengue viral illness, the input of fluid is typically greater than output seen, so the fluid input/output ratio cannot be used to guide the fluid resuscitation therapy.

The targets for fluid resuscitation in these patients are:

- Improving peripheral and central circulation. This is seen by reducing tachycardia, improved pulse volume, blood pressure, and warm extremities, <2 s capillary refill time.

**TABLE 10.5** Dengue viral illness Group B management algorithm.

| | Dengue viral illness with warning signs | |
|---|---|---|
| **Classification** | **Group B (inpatient care)** | |
| **Criteria** | Patients with any of the following:<br>• Coexisting conditions like diabetes mellitus, old age, pregnancy, and infancy<br>• Social: Living alone, house far from hospital | Presence of warning signs:<br>• Abdominal pain or tenderness<br>• Persistent vomiting<br>• Clinical fluid accumulation<br>• Mucosal bleeding<br>• Lethargy/restlessness<br>• Liver enlargement >2 cm<br>• Increase in hematocrit |
| **Lab tests** | • Full blood count including hematocrit | |
| **Treatment** | • Encourage fluid intake<br>• If not possible then intravenous fluid therapy in the form of 0.9% saline or Ringer lactate | • Baseline hematocrit before intravenous fluid therapy<br>• 0.9% saline or Ringer lactate: 5–7 mL/kg/h for 1–2 h, then reduce to 3–5 mL/kg/h for 2–4 h, and then reduce to 2–3 mL/kg/h or less based on clinical response<br>• Recheck clinical status and repeat hematocrit<br>  • Minimal increase in hematocrit: Continue fluids at 2–3 mL/kg/h for 2–4 h<br>  • Rapid increase in hematocrit: fluid increase to 5–10 mL/kg/h for 1–2 h<br>  • If the hematocrit levels are decreasing then the intravenous fluid rate is gradually decreased. |
| **Monitoring** | • Temperature<br>• Fluid intake and output<br>• Urine output<br>• Hematocrit | • Peripheral perfusion and vital signs every 1–4 h<br>• Urine output every 4 h<br>• Hematocrit: Before and after fluid therapy, followed by 6–12 hourly<br>• Blood glucose<br>• Organ functions tests as needed |

Adapted from World Health Organization: Handbook for Clinical Management of Dengue (2012).

- Improvement in end-organ perfusion in the patient. This is seen by improvement in the conscious level of the patients. These patients typically become more alert and less restless as the end-organ perfusion improves. Urine output improves as well to >0.5 mL/kg/hour and the metabolic acidosis also improves.

## Treatment of compensated shock

The plan of action in these patients is:

- Hematocrit level is obtained to get a reference point before starting the intravenous fluid therapy. After this, isotonic crystalloid solution is started at the rate of 5 mL/kg/h over a period of 1 h in adults. In infants and children, the fluid rate is kept at 10–20 mL/kg/h over a period of 1 h. After the initial fluid resuscitation, the patient's condition is reassessed. Vital signs, hematocrit, capillary refill time, and urine output are recorded.
- **Adult patient**: as the condition improves the fluid replacement rate is reduced gradually. Initially it is reduced to 5–7 mL/kg/hour over a period of 1–2 h, then it is reduced further to 3–5 mL/kg/hour over a period of 2–4 h, and then finally it should be reduced to 2–3 mL/kg/hour maintained over 24–48 h. As the patient improves transition to oral fluids. The upper limit of intravenous fluid use should not exceed 48 h.
- **Infants/children**: as the patient improves, the fluid replacement rate is reduced gradually. Initially, the rate is reduced to 10 mL/kg/h over a period of 1–2 h, then it is further reduced to 7 mL/kg/h over a period of 2 h, and then finally it is reduced to 3 mL/kg/h maintained over a period of 24–48 h. As the patient improves transition to oral fluids. The upper limit of intravenous fluid use should not exceed 48 h.

If the patient is still in a state of shock and the vitals remain unstable, the hematocrit should be reassessed after the first bolus of intravenous fluid bolus. The following measures should be taken:

- **Adults**: If the hematocrit remains high i.e., >50% or rises, a second bolus of intravenous fluid should be repeated using a crystalloid/colloid solution at a rate of 10–20 mL/kg/hour over a period of 1 h. If an improvement is seen after a second bolus in the condition of the patient then the fluid should be reduced to a rate of 7–10 mL/kg/hour over a period of 1–2 h. Further reduction is done as stated above.

If the hematocrit level is reduced when compared to the initial reference value, and the patient continues to be unstable (vital signs unstable), then a vigilant watch should be kept for active bleeding. The patient should be immediately cross-matched and transfused with fresh whole blood or fresh packed red cells in a case of severe bleeding. If no active source of bleeding is seen, another bolus of 10–20 mL of colloid should be given. The patient is then reassessed both clinically and for hematocrit levels.

- **Infants/children**: If the hematocrit levels remain high or rise, the fluid replacement should be switched to a colloid solution and started at a rate of 10–20 mL/kg/hour. After the first bolus it should be gradually reduced to 10 mL/kg/hour over a period of 1 h and then reduced further to a rate of

7 mL/kg/h. When the patient starts to show improvement, the fluid should be changed to a crystalloid-based solution.

If the hematocrit level is reduced when compared to the initial reference value, and the patient continues to be unstable (vital signs unstable), then a vigilant watch should be kept for bleeding. The patient should be immediately cross-matched and transfused with fresh whole blood or fresh packed red cells in a case of severe bleeding. If no active source of bleeding is seen, another bolus of 10–20 mL of colloid over 1 h should be given. The patient is then reassessed using both clinical parameters and serial hematocrit levels (Figs. 10.1–10.3).

## *Treatment of profound hypotensive shock (undetectable blood pressure and pulse)*

In profound hypotensive shock due to Dengue viral illness, the blood pressure cannot be measured accurately using a cuff and the pulse is very difficult to detect. Any patient suffering from profound hypotensive shock should be managed aggressively. If the decreased peripheral perfusion is left untreated, it

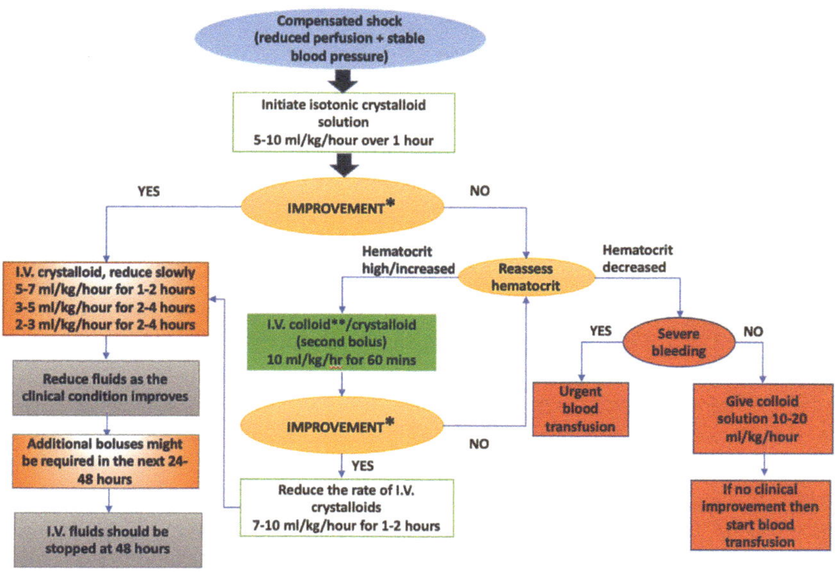


FIGURE 10.1 Algorithm for fluid management of compensated shock: in adults. *Reevaluate the patient's clinical condition including vital signs, temperature of extremities, pulse volume, and capillary refill time; **Colloid is preferred if the patient has already received prior boluses of crystalloid; *I.V.*, Intravenous. *Adapted from World Health Organization: Handbook for Clinical Management of Dengue (2012).*

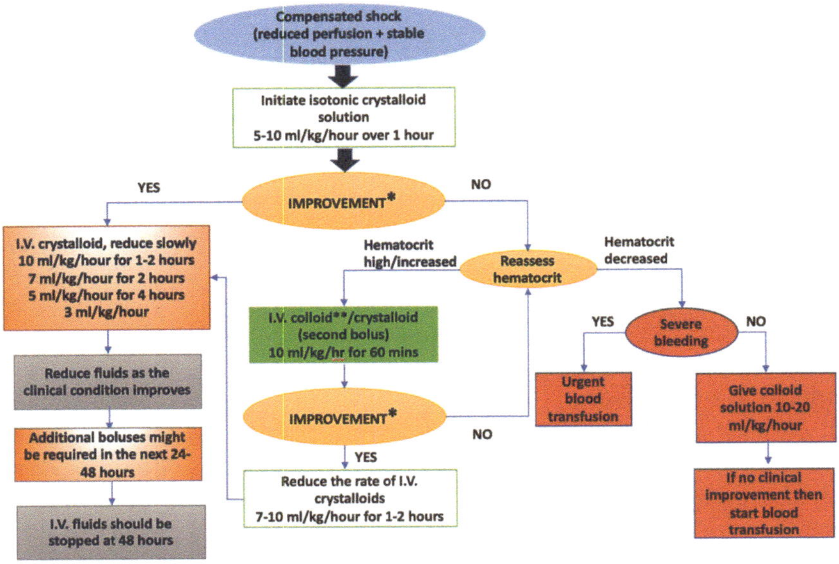


FIGURE 10.2 Algorithm for fluid management of compensated shock: in infants/children. *Reevaluate the patient's clinical condition including vital signs, temperature of extremities, pulse volume, and capillary refill time; **Colloid is preferred if the patient has already received prior boluses of crystalloid; *I.V.*, Intravenous. *Adapted from World Health Organization: Handbook for Clinical Management of Dengue (2012).*

can lead to severe ischemic injury and result in multi-organ dysfunction. As the hypovolemic shock develops it leads to decreased volume status and a significant drop in the systolic blood pressure. As a result, the vital organs are unable to meet the oxygen demand due to decreased oxygen supply. This further leads to lactic acidosis as the cells of the organs shift from aerobic metabolism to anaerobic metabolism. This is further worsened due to diverted blood supply to the vital organs such as brain and heart. If left untreated it propagates into more ischemia, lactic acidosis, and even death.

The first step is to start intravenous fluid resuscitation using colloid or crystalloid solution at the rate of 20 mL/kg given as a bolus over a period of 15–30 min. This is done to bring the patient out of shock as fast as possible. In patients who have a very low pulse pressure (<10 mmHg), blood pressure needs to be restored as urgently as possible. In these cases, colloids are the preferred fluid of choice. If peripheral venous access cannot be taken then intraosseous route should be explored. Studies have shown that colloids work better than crystalloids in a state of intractable shock [3–5]. When treating hypotensive shock the major concern is treating hypovolemia rapidly without causing volume overload. The fluid volume used in the resuscitation efforts depends on the ability of the fluid to remain in the intravascular space and

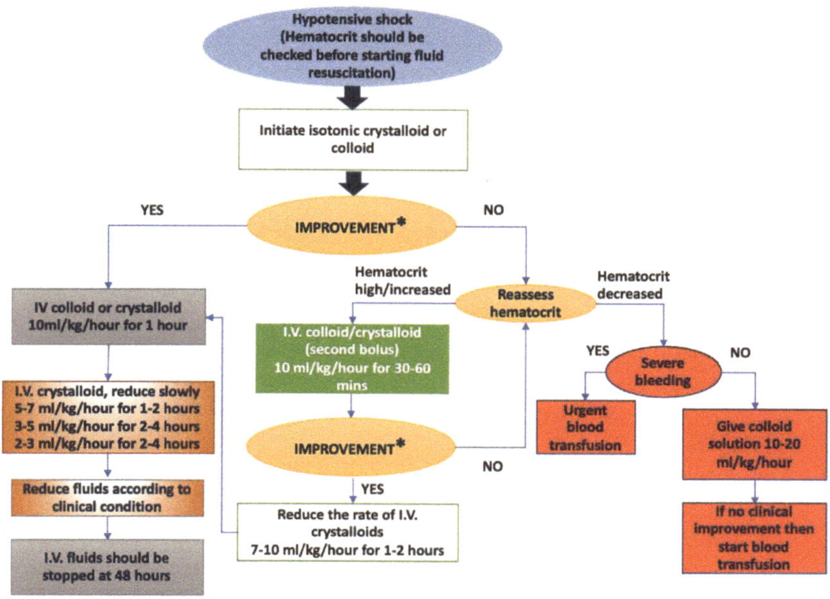


FIGURE 10.3 Algorithm for fluid management of hypotensive shock: in infants/children and adults. *Reevaluate the patient's clinical condition including vital signs, temperature of extremities, pulse volume, and capillary refill time; **Colloid is preferred if the patient has already received prior boluses of crystalloid; *I.V.*, intravenous. *Adapted from World Health Organization: Handbook for Clinical Management of Dengue (2012).*

creating an osmotic gradient [6]. With an increased osmotic gradient, more fluid moves into the intravascular compartment thereby improving the hemodynamic stability of the patient. Crystalloids remain in the intravascular compartment for a shorter duration than colloids. They may also lead to higher chance of pulmonary edema especially when the pulmonary circulation is affected by the increased systemic capillary permeability. Colloids are believed to be better than crystalloids in having lesser risk of fluid overload and therefore pulmonary edema. According to one prospective study conducted in trauma patients, colloids were faster than crystalloids to stabilize hemodynamic profile of the patient [7]. This study also showed that fluid required for resuscitation was lesser in colloids than crystalloids. However, one meta-analysis has shown that colloids lead to increased mortality rates when used in critically ill patients [8].

After the bolus, the condition of the patient should be assessed. The following measures are taken based on the condition of the patient:

- **Adults**: if the condition improves then colloid/crystalloid infusion is started at the rate of 10 mL/kg/hour over a period of 1 h. After this the fluid is switched to a crystalloid-based solution. The fluid rate is gradually reduced to a rate of 5–7 mL/kg/hour over a period of 1–2 h, then reduce it to a rate of 3–5 mL/kg/hour over a period of 2–4 h, and then finally reduce to a maintenance rate of 2–3 mL/kg/hour and lesser over a period of 24–48 h. As the urine output improves and the oral fluid intake improves, consider reducing the intravenous fluid even further. The maximum duration of intravenous fluid intake should not exceed 48 h.

- **Infants/children**: initially these patients are started on a colloid infusion rate of 10 mL/g/h over a period of 1 h. After this the fluid is switched to a crystalloid infusion. The fluid rate should be gradually reduced to a rate 10 mL/kg/hour over a period of 1 h, then reduced to a rate of 7.5 mL/kg/hour over a period of 2 h, then to a rate of 5 mL/kg/hour over a period of 4 h and then a maintenance rate of 3 mL/kg/h over a period of 24–48 h should be adopted. As the urine output improves and the oral fluid intake improves, consider reducing the intravenous fluid even further. The maximum duration of intravenous fluid intake should not exceed 48 h.
- **Unstable patients**: if the vital signs are unstable and the patient still remains in a state of shock, hematocrit should be reviewed again.
  - **Hematocrit normal to low** (<30%–35% in infants, <35%–40% in children and adult females, <40%–45% in adult males)**:** this can be sign of active bleeding. The patient should be immediately cross-matched and transfused with fresh whole blood or fresh packed red cells in a case of severe bleeding. If no active source of bleeding is seen, another bolus of 10–20 mL of colloid over 30 min to 1 h should be given. The patient is then reassessed both clinically and for hematocrit levels.
  - **Hematocrit high:** in this scenario, the intravenous fluid should be switched to a colloid solution and a second bolus should be given at the rate of 10–20 mL/kg over a period of 30 min to 1 h. After reassessing the condition of the patient if the condition improves, the rate of fluids is reduced to 7–10 mL/kg/h over a period of 1–2 h. After this the fluids are switched back to a crystalloid solution. Further reduction in the infusion rate is done as stated above.
- After the second bolus, if the patient still remains unstable, the hematocrit needs to be repeated.
  - **Hematocrit decreased** (from the previous value): This is a sign of bleeding. The patient should be immediately cross-matched and transfused with fresh whole blood or fresh packed red cells.
  - **Hematocrit increased** (from the previous value): The patient should be given a third bolus of colloid solution at a rate of 10–20 mL/kg over a period of 1 h. The rate is then reduced to 7–10 mL/kg/hour over a period of 1–2 h. If the patient improves, the fluid is then switched back to a crystalloid solution and the rate is reduced as mentioned above. Another hematocrit needs to be taken after the third bolus if the patient remains unstable.

- During the next 24 h more fluid boluses might be required. The volume and rate of these need to be decided based on the clinical response of the patient. These patients suffering from severe Dengue viral illness need to be admitted in an intensive care unit.

These patients need to be monitored very closely. Input and output fluid charts should be maintained in full detail. An arterial line should be used for the estimation of blood pressure in these shock patients as the cuff measures might not be very accurate. A continuous bladder catheter can be used to record urine output accurately. The patients should be additionally monitored for arterial or venous or capillary blood gases, blood glucose, total carbon dioxide/bicarbonate, lactate, liver profile, coagulation profile, renal profile etc (Table 10.6).

## Management of bleeding

Bleeding (mucosal) is one of the most common symptom of Dengue viral illness. As long as the patient remains hemodynamically stable with the fluid resuscitation, bleeding may not be a major issue. As the patient recovers, the mucosal bleeding usually subsides. Patients should be advised strict bed rest and trauma protection if they have severe thrombocytopenia. Intramuscular injections are to be avoided in these patients.

Major bleeding risk factors include the folllowing:

- Profound shock
- Refractory shock
- Hypotensive shock
- Nonsteroidal anti-inflammatory agents (NSAIDS)
- Anticoagulant therapy
- Metabolic acidosis
- Peptic ulcer disease
- Intramuscular injection
- Trauma

Patients with profound/prolonged shock might be suffering from unrecognized severe bleeding [9]. Hematocrit is initially very high due to plasma leakage. When there is severe bleeding, the level of hematocrit will decrease from this high level. As the initial hematocrit level is high, severe bleeding might bring it to a normal level or low normal level after fluid replacement therapy. Special care should be taken to look for severe bleeding in these patients.

Conditions in which severe bleeding should be watched and treated immediately include the following:

- Unstable hemodynamic status with overt/persistent bleeding regardless of the hematocrit level.

**TABLE 10.6 Dengue viral illness Group C management algorithm.**

| | Severe Dengue viral illness |
|---|---|
| **Classification** | **Group C (emergency treatment)** |
| **Criteria** | Patients with any of the following:<br>• Severe plasma leakage<br>• Respiratory distress due to fluid accumulation<br>• Bleeding<br>• Organ impairment |
| **Lab tests** | • Full blood count<br>• Hematocrit<br>• Organ function tests if needed such as renal function test, and/or liver function test |
| **Treatment** | **Treatment of compensated shock:**<br>• Intravenous fluid resuscitation in the form of isotonic crystalloid at 5–10 mL/kg/hour over 1 h Reassess patient<br>• If patient improves then reduce intravenous fluids gradually from 5 to 7 mL/kg/hour for 1–2 h; then reduce to 3–5 mL/kg/hour for 2–4 h; then reduce to 2–3 mL/kg/hour for 2–4 h; then reduced to lower rates depending on hemodynamic status of the patient<br>• If the patient is still unstable then give a fluid bolus Check hematocrit<br>• **Hematocrit high:** give another fluid bolus of crystalloid solution at 10–20 mL/kg/h for 1 h. As the patient improves reduce rate to 7–10 mL/kg hour for 1–2 h and then reduce further as above<br>• **Hematocrit low:** this can indicate active bleeding. Patient should be cross-matched and blood transfusion should be started immediately<br>• **Treatment of hypotensive shock:** Intravenous fluid resuscitation in the form of crystalloid/colloid at 20 mL/kg over 15 min.<br>• If patient improves then crystalloid/colloid solution is given at 10 mL/kg/hour over 1 h. Later the fluid rate is reduced gradually as above. If the patient is still unstable then review the hematocrit before the first bolus.<br>• **Hematocrit low** (<40% in children and adult females, <45% in adult males): this is a sign of active bleed, patient needs to be cross-matched and transfused.<br>• **Hematocrit high** (compared to baseline value): Change the intravenous fluid to colloids at 10–20 mL/kg as a second bolus over 30 min to 1 h.<br>• Reassess the patient after second bolus.<br>• If the patient **improves**, decrease the rate to 7–10 mL/kg/hour for 1–2 h. After this the intravenous fluids are switched back to crystalloid and then gradually reduced as above.<br>• **Hematocrit decreases:** Sign of bleeding, this is a sign of active bleed, patient needs to be cross-matched and transfused. |

**TABLE 10.6 Dengue viral illness Group C management algorithm.—cont'd**

| | Severe Dengue viral illness |
|---|---|
| Classification | Group C (emergency treatment) |
| | • **Hematocrit increases or** remains **high** (>50%): Continue colloid infusion at 10–20 mL/kg as a third bolus over 1 h; reduce the rate to 7–10 mL/kg/hour for 1–2 h; then the fluid is changed back to crystalloid solution and reduced as above.<br>• **Treatment of hemorrhagic complications:** 5–10 mL/kg of fresh packed red cells or 10–20 mL/kg fresh whole blood |

Adapted from World Health Organization: Handbook for Clinical Management of Dengue (2012).

- Unstable hemodynamic status even after fluid boluses along with a decreased hematocrit.
- Unresponsive refractory shock even after high volume fluid resuscitation (40–60 mL/kg).
- Low/normal hematocrit in the presence of hypotensive shock.
- Patients presenting with severe abdominal tenderness/distension and persistent metabolic acidosis. Blood pressure can be well maintained in these patients.

Treatment of choice is blood transfusion after cross matching the patient. There is an increased risk of fluid overload in these patients if blood transfusion is done. It is important to not wait for hematocrit to decrease too low before initiating blood transfusion.

Treatment of complications due to hemorrhage include:

- If the source of bleeding can be identified then the initial step should be to try to stop the bleeding. In case of epistaxis, nasal packing with intranasal adrenaline should be tried.
- The amount of blood transfusion should be matched to the amount of blood loss if quantifiable.
- If not then 5–10 mL/kg of packed red cells or 10–20 mL/kg fresh whole blood.
- Blood transfusion should be repeated if the blood loss continues or the hematocrit level does not improve.
- Platelet transfusion has not been proven to prevent hemorrhage in those with severe thrombocytopenia ($<10,000/mm^3$) unless patient is actively bleeding [10,11].
- Platelet transfusion can however be used if obstetrical delivery or any other surgical intervention is planned in these patients [12].

- H-2 antagonist and proton pump inhibitors (PPI) have been used in the event of gastrointestinal bleeding [13].
- Extra care should be taken to prevent trauma when inserting bladder catheters, nasogastric tubes, and central venous catheters.
- Blood transfusion is only advised in the event of a severe bleeding episode.
- Unecessary blood transfusion can raise the hematocrit and give a false impression of a hemo-concentrated state. This can lead to unwarranted use of fluid therapy and thereby fluid overload.

## Vaccines

Mosquito control and vaccine development are two major approaches for prevention of Dengue viral illness in endemic areas. Feasibility of developing an effective vaccine against Dengue virus emerged due to the fact that infection with Dengue viral illness provides long-term protection against disease caused by reinfection with that particular sero type, and also short-lived cross-protection to the other three Dengue virus types. A number of vaccine candidates are in development against Dengue virus including; live-attenuated and inactivated viruses, recombinant proteins, DNA vaccines, and also heterologous prime/boost strategies are under development [14]. Few live-attenuated vaccine candidates are currently being tested in phase III trials including DENVax, Inviragen, and TV003/TV005 [15,16]. The major hurdles in development of Dengue virus vaccines are the lack of the best animal model, various immune status of individual especially in endemic areas, no defined cut off of protective immunity, and various strains of Dengue virus. Development of polyvalent vaccine against Dengue virus types makes sense but poses several challenges including preservatives/buffers used for one antigen may alter the potency of other antigens, titers of antibodies in patients may be lower when polyvalent vaccines are administered compared to the equivalent monovalent versions in separate administrations, and manufacturers must demonstrate that there are no differences in physical, chemical, and immunological responses for simultaneous administration of multiple antigens in comparison to the individual antigens separately [12,13]. Development of a polyvalent vaccine against Dengue virus has been hindered by antibody-dependent enhancement of disease by induction of suboptimal levels of neutralizing antibodies which poses the risk of enhanced disease during natural infection [14]. The tetravalent live-attenuated Dengue virus vaccine showed differential immunogenicity and efficacy among the serotypes and may be associated with enhanced disease in some subjects, which in turn is related to imbalanced infectivity [15,16]. Several research and development efforts are ongoing to find a better effective and accessible vaccine to prevent people at risk from having such a fatal disease.

CYD-TDV (Dengvaxia), four chimeric yellow fever 17D-Dengue virus vaccine, has recently been approved by FDA in May 2019 with caution that it should only be used in individuals with previous Dengue viral illness (laboratory-confirmed) and living in endemic areas [17]. This vaccine has already been previously licensed and used in several countries in Latin America, South East Asia, and Europe. CYD-TDV has premembrane and envelope proteins from each of the four Dengue virus types which replace the same proteins in a yellow fever 17D backbone virus. Vaccine is indicated in children $\geq$9 years and adolescents $\leq$16 years with subQ dose of 0.5 mL/dose for three doses administered at 0, 6, and 12 months. No dose adjustment is needed for hepatic or renal impairment. Vaccine is not intended for use in adults >16 years of age, those who are immunocompromised, and those not previously infected with Dengue viral illness.

The licensure and efficacy of CYD-TDV was based on two large phase III randomized controlled trials [18,19]. In both trials, the CYD-TDV vaccine was administered in three doses at months 0, 6, and 12. Capeding et al. [18] enrolled children aged 2 to 14 in five countries in the Asia–Pacific region; and the other study, Villar et al. [19], enrolled children aged 9–16 years in Latin America and the Caribbean Islands. Both of these trials showed efficacy of vaccine against virologically confirmed disease of any severity caused by any Dengue virus type; relative risk reduction was 57% and 61%, respectively. Vaccine efficacy was found to be 80%–95% against Dengue viral illness hemorrhagic fever or requiring hospitalization. Dengue virus-3 and Dengue virus-4 were found to be most protected by vaccine i.e., approximately 75%, as compared to Dengue virus-1 (50%) and Dengue virus-2 (35%–42%). Age also played an important determinant in measuring the efficacy of vaccine, especially being least effective in children who are 2–5 years (34%–36%), and the same age group had more frequent hospitalizations. Children who did not have detectable Dengue virus-neutralizing antibodies prior to vaccination were also found to be least protected by vaccine (relative risk reduction of 34%–36%). The safety profile was considered good, and there was no indication of more severe Dengue viral illness in breakthrough cases in vaccine recipients that occurred over the 25 months of active case surveillance. Sridhar et al. [5] performed post hoc analysis on safety and efficacy of vaccine using anti-nonstructural protein 1 (NS1) IgG enzyme-linked immunosorbent assay. This case cohort study reported that vaccine was protective against severe virologically confirmed Dengue viral illness (0.75 vs. 2.47%) and hospitalization (3.06 vs. 1.87%) for 5 years in persons who had exposure to Dengue virus before vaccination as compared to those who did not. Hadinegoro et al. [20] reported the results of long-term follow-up interim analyses and efficacy analyses on CYD 14, CYD 15, and CYD23/57, and found that there was unexplained higher incidence of hospitalization for Dengue viral illness in year three among children younger than 9 years of age with pooled risk of 1.58

(95% confidence interval, 0.83 to 3.02) as compared to 0.50 (95% confidence interval, 0.29 to 0.86) among those 9 years of age or older. The overall risk among children 2–16 years of age was lower in the vaccine group than in the control group 0.84 (95% confidence interval, 0.56 to 1.24). Rosa et al. [21] performed a systemic review and meta-analysis on efficacy, immunogenicity, and safety of CYD-TDV vaccine in children 2–17 years of age, and found that vaccine is considered safe and able to partially protect children and adolescents against four serotypes of Dengue virus for a 1-year period, and most protective against Dengue virus-3 (relative risk reduction of 75%) and Dengue virus-4 (relative risk reduction of 77%). Clinical development program of CYD-TDV (Dengvaxia) includes 31 clinical studies, which are 5 Phase I, 17 Phase II, and 9 Phase III (Table 10.4).

In 2017, the vaccine manufacturer published a report that the vaccine showed the persistent beneficial effect in 6 years of clinical data in individuals who had been previously infected with Dengue viral illness prior to vaccination. In individuals with no prior episode of Dengue viral illness, however, vaccination was associated with an increased risk of severe disease and hospitalization. Therefore, the manufacturer updated the label, advising that individuals with no prior Dengue viral illness avoid vaccination [20]. In December 2017, the World Health Organization issued a statement based on above mentioned vaccine manufacturer report indicating that the vaccine is protective against severe Dengue viral illness for individuals with Dengue virus seropositivity at the time of first vaccination but that the risk of severe Dengue viral illness is significantly increased for individuals with Dengue virus seronegativity at the time of first vaccination [21]. In April 2018, the World Health Organization advised that the vaccine should not be used until prior Dengue viral illness can be confirmed at the time of administration [20,22].

The mechanisms for CYD-TDV vaccine efficacy in seropositive individuals and the increased risk of severe Dengue viral illness in seronegative individuals are uncertain. Only a subset of individuals underwent scheduled serologic monitoring prior to and following vaccination, so analyses of the immune profiles associated with various outcomes are limited. A statistical analysis found that higher postvaccination titers of neutralizing antibodies were associated with a significantly higher vaccine efficacy for all Dengue virus types, baseline serostatus groups, and age groups. However, it is believed that the relationship between vaccine-induced immune responses and efficacy is likely more complex than explained by any one single factor, because the increase in titers was not significantly associated with vaccine efficacy, and no absolute correlate of protection could be identified. The data suggests that neutralizing antibodies are at best a relative correlate of protection [23].

This antibody augmentation phenomenon led to "Dengvaxia controversy or Dengvaxia mess" in Philippines, where several deaths (133 children and 3

**TABLE 10.7** Clinical development program of CYD-TDV (Dengvaxia).

| Study identifier | Study design | Test product | Trial period | Status | Number of individuals | Outcome |
|---|---|---|---|---|---|---|
| CYD01 | Phase I, randomized | ChimeriVax$^{TM}$-DEN2 | Mar 2002 –Jun 2002 | Completed | 56 | *Safety and immunogenicity profile of the vaccine is consistent with that of YF-VAX<br>*Preimmunity to Yellow fever virus does not interfere with ChimeriVax-DEN2 immunization, but induces a long lasting and cross neutralizing antibody response to all 4 DEN serotypes |
| CYD02 | Phase I, randomized | Dengvaxia (~4 $log_{10}CCID_{50}$/serotype 1, 2, 3, 4) | 2003–04 | Completed | 99 | NA |
| CYD04 | Phase I, randomized | Dengvaxia (~5 $log_{10}CCID_{50}$/serotype 1, 2, 3, 4) | 2005–07 | Completed | 66 | TDV was well-tolerated and induced full seroconversion against all WHO reference strain serotypes after 3 doses |
| CYD05 | Phase I, randomized | Dengvaxia (~5 $log_{10}CCID_{50}$/serotype 1, 2, 3, 4) | 2006–2012 | Completed | 126 | Vaccine regimen of either three TDV vaccinations administered over a year or two TDV vaccinations given more than 8 months apart resulted in a balanced antibody response to all four Dengue virus serotypes in flavivirus-exposed population, including children |

*Continued*

**TABLE 10.7** Clinical development program of CYD-TDV (Dengvaxia).—cont'd

| Study identifier | Study design | Test product | Trial period | Status | Number of individuals | Outcome |
|---|---|---|---|---|---|---|
| CYD06 | Phase I, randomized | Dengvaxia (~5 $log_{10}CCID_{50}$/serotype 1, 2, 3, 4) | 2006–07 | Completed | 126 | Three-dose TDV regimen had a favorable safety profile in children and adults and elicited neutralizing antibody responses against all 4 serotypes |
| CYD08 | Phase II, randomized | Dengvaxia (~5 $log_{10}CCID_{50}$/serotype 1, 2, 3, 4) | 2010–12 | Completed | 222 | CYD-TDV has an acceptable safety and immunogenicity profile in toddlers and when coadministered with MMR |
| CYD10 | Phase IIa, monocenter, controlled, open trial | Dengvaxia (~5 $log_{10}CCID_{50}$/serotype 1, 2, 3, 4) | 2006–07 | Completed | 35 | Prior exposure to YF or monovalent Dengue virus vaccines had no adverse effects on the safety or incidence of viremia associated with TDV, but it increased immunogenicity |
| CYD11 | Phase IIa, randomized | Bivalent or tetravalent Dengvaxia (~5 $log_{10}CCID_{50}$/serotype) Blending tetravalent Dengvaxia (~5 $log_{10}CCID_{50}$/serotype 1, 3, 4) + Vero Dengue vaccine (~4 $log_{10}CCID_{50}$/serotype 2) | 2008–09 | Completed | 155 | TDV in the Japanese encephalitis primed group were in general higher than after the first injection in the other groups. All tested regimens were well-tolerated without marked differences between groups. Bivalent vaccination showed no advantage in terms of immunogenicity |

| | | | | | | |
|---|---|---|---|---|---|---|
| CYD12 | Phase II, randomized | Dengvaxia formulations: 5555 ($\sim$5 $log_{10}CCID_{50}$/serotype 1, 2, 3, 4) 5553 ($\sim$5 $log_{10}CCID_{50}$/serotype 1, 2, 3) and ($\sim$3 $log_{10}CCID_{50}$/serotype 4) 4444 ($\sim$4 $log_{10}CCID_{50}$/serotype 1, 2, 3, 4) | 2008–09 | Completed | 260 | Reducing the dose of serotype 4 antigen (5553 formulation) creates an imbalance in the immune response to CYD-TDV. Immune responses to CYD-TDV 5555 were slightly higher than to the 4444 formulation. Development of CYD-TDV 5555 has subsequently been pursued |
| CYD13 | Phase II, randomized | Dengvaxia ($\sim$5 $log_{10}CCID_{50}$/serotype 1, 2, 3, 4) | 2009–11 | Completed | 600 | CYD-TDV had a favorable safety profile and elicited antibody responses against all 4 Dengue virus serotypes in 9–16 year olds in Latin America. These findings support the continued development of CYD-TDV |
| CYD14 | Phase III, randomized | Dengvaxia ($\sim$5 $log_{10}CCID_{50}$/serotype 1, 2, 3, 4) | 2011 | Completed | 10,275 | Vaccination could reduce the incidence of symptomatic infection and hospital admission and has the potential to provide an important public health benefit |
| CYD15 | Phase III, randomized | Dengvaxia ($\sim$5 $log_{10}CCID_{50}$/serotype 1, 2, 3, 4) | 2011–12 | Completed | 20,869 | CYD-TDV Dengue virus vaccine was efficacious against virologically confirmed disease and severe virologically confirmed disease, and led to fewer hospitalizations for in five Latin American countries where Dengue viral illness is endemic |

*Continued*

**TABLE 10.7 Clinical development program of CYD-TDV (Dengvaxia).—cont'd**

| Study identifier | Study design | Test product | Trial period | Status | Number of individuals | Outcome |
|---|---|---|---|---|---|---|
| CYD17 | Phase III, randomized | Dengvaxia (~5 $\log_{10}CCID_{50}$/serotype 1, 2, 3, 4) | 2010–12 | Completed | 715 | For serotype 2, although equivalence was not shown between two lots, the geometric mean neutralizing antibody titers observed in the phase III lots were consistently higher than those for the phase II lot. As such, in our view, biological equivalence for all serotypes was demonstrated |
| CYD22 | Phase II, randomized | Dengvaxia (~5 $\log_{10}CCID_{50}$/serotype 1, 2, 3, 4) | 2009–14 | Completed | 252 | CYD-TDV demonstrated long-term immunogenicity and was well-tolerated for 4 years after the third vaccine dose Dengue |
| CYD23 | Phase IIb, randomized | Dengvaxia (~5 $\log_{10}CCID_{50}$/serotype 1, 2, 3, 4) | 2009–12 | Completed | 4002 | Data shows for the first time that a safe vaccine against virus is possible. Ongoing large-scale phase 3 studies in various epidemiological settings will provide pivotal data for the CYD Dengue virus vaccine candidate |
| CYD57 | Monocenter, safety follow-up study of CYD23 | No vaccine administration | 2013–16 | Completed | 3203 | Safe vaccine against Dengue virus is possible |

| | | | | | | |
|---|---|---|---|---|---|---|
| CYD24 | Phase II, randomized | Dengvaxia (~5 $log_{10}CCID_{50}$/serotype 1, 2, 3, 4) | 2008–10 | Completed | 300 | 3-Dose regimen of CYD-TDV had a good safety profile in 2–11 year olds with a history of Yellow fever vaccination and elicited robust antibody responses that were balanced against the four serotypes |
| CYD28 | Phase II, randomized | Dengvaxia (~5 $log_{10}CCID_{50}$/serotype 1, 2, 3, 4) | 2009–14 | Completed | 1198 | Postdose 3%, 66.5% of all participants were seropositive to all four serotypes, and 87.2% were seropositive to $\geq 3$ serotypes; geometric mean neutralizing antibody titers for all participants ranged from 43.0 against Dengue virus serotype 1 to 100 against Dengue virus serotype 4. GMTs were higher in children than in adolescents |
| CYD29 | Phase III, randomized | Dengvaxia (~5 $log_{10}CCID_{50}$/serotype 1, 2, 3, 4) | 2011–13 | Completed | 792 | Concomitant administration of yellow fever vaccine with CYD-TDV has no relevant impact on the immunogenicity or safety profile of the Yellow fever vaccine |
| CYD30 | Phase II, randomized | Dengvaxia (~5 $log_{10}CCID_{50}$/serotype 1, 2, 3, 4) | 2010–12 | Completed | 150 | CYD-TDV vaccination elicited a neutralizing antibody response against serotypes 1–4 and was well-tolerated in children/adolescents in a Dengue viral illness-endemic region |
| CYD32 | Phase III, randomized | Dengvaxia (~5 $log_{10}CCID_{50}$/serotype 1, 2, 3, 4) | 2010–12 | Completed | 250 | Satisfactory safety profile and a balanced humoral immune response against all four DENV serotypes for CYD-TDV administered via a three-dose regimen to children in Malaysia |

*Continued*

**TABLE 10.7 Clinical development program of CYD-TDV (Dengvaxia).—cont'd**

| Study identifier | Study design | Test product | Trial period | Status | Number of individuals | Outcome |
|---|---|---|---|---|---|---|
| CYD33 | Phase III, randomized | Dengvaxia (~5 $log_{10}CCID_{50}$/serotype 1, 2, 3, 4) | 2011–14 | Completed | 720 | No difference was observed on the CYD-TDV profile when administered concomitantly or sequentially with the DTaP-IPV//Hib booster vaccine |
| CYD47 | Phase II, randomized | Dengvaxia (~5 $log_{10}CCID_{50}$/serotype 1, 2, 3, 4) | 2012–13 | Completed | 189 | The immunogenicity and safety of CYD-TDV was satisfactory in both Dengue virus seropositive and seronegative Indian adults |
| CYD51 | Phase II, randomized | Dengvaxia (~5 $log_{10}CCID_{50}$/serotype 1, 2, 3, 4) | 2011–13 | Completed | 390 | Live-attenuated CYD-TDV vaccine given in a compressed schedule in a nonendemic setting can elicit similar antisbody responses to the licensed CYD-TDV schedule |
| CYD63 | Phase II, randomized | Dengvaxia (~5 $log_{10}CCID_{50}$/serotype 1, 2, 3, 4) | 2016-Till date | Ongoing | 118 | NA |
| CYD64 | Phase II, randomized | Dengvaxia (~5 $log_{10}CCID_{50}$/serotype 1, 2, 3, 4) | 2016-Till date | Ongoing | 251 | NA |
| CYD65 | Phase II, randomized | Dengvaxia (~5 $log_{10}CCID_{50}$/serotype 1, 2, 3, 4) | 2016-Till date | Ongoing | 1050 | NA |

adults as of June 3, 2019) have been allegedly attributed to complication from Dengvaxia administration [24]. In April 2016, about 700,000 individuals in Philippines received at least one dose of the vaccine as part of the Dengue vaccination campaign launched by Department of Health. Vaccine manufacturer, in November 2017, made a statement that CYD-TDV poses higher risks to people without prior Dengue viral infection, which led to temporarily suspension of school-based Dengue virus vaccination [22]. In April 2018, Public Attorney's Office filed the criminal charges (reckless imprudence resulting in homicide under Article 365 of the Revised Penal Code and violation of Republic Act No. 9745 [Anti-Torture Act] and torture) against the Health Secretary and other former officials. Parents of children whose deaths had been linked to Dengvaxia have also filed the charges against several health officials. In March 2019, The Philippine Department of Justice announced that it had found probable cause to indict officials from vaccine manufacturer and Philippine health officials over deaths linked to use of a Dengvaxia [24,25].

Overall, CYD-TDV demonstrated superior efficacy and clinical benefit in seropositive recepients compared with seronegative vaccine recipients. However, the only FDA-approved vaccine, CYD-TDV, has several limitations including lower efficacy was observed in children with <9 years and >16 years old, higher risk of hospitalization, and severe Dengue viral illness in children 2–5 years old, lower efficacy against DENV2 serotypes, and protective efficacy also depended on prior Dengue viral illness serostatus. Therefore, the quest of a better vaccine candidate continues (Table 10.7).

## References

[1] Harris E, et al. Fluid intake and decreased risk for hospitalization for dengue fever, Nicaragua. Emerg Infect Dis 2003;9(8):1003–6.

[2] Kramer MS, et al. Risks and benefits of paracetamol antipyresis in young children with fever of presumed viral origin. Lancet 1991;337(8741):591–4.

[3] Ngo NT, et al. Acute management of dengue shock syndrome: a randomized double-blind comparison of 4 intravenous fluid regimens in the first hour. Clin Infect Dis 2001;32(2):204–13.

[4] Dung NM, et al. Fluid replacement in dengue shock syndrome: a randomized, double-blind comparison of four intravenous-fluid regimens. Clin Infect Dis 1999;29(4):787–94.

[5] Wills BA, et al. Comparison of three fluid solutions for resuscitation in dengue shock syndrome. N Engl J Med 2005;353(9):877–89.

[6] Pacheco Jr AM, et al. Hypertonic volume therapy: feasibility in the prevention and treatment of multiple organ failure and sepsis. Sao Paulo Med J 1995;113:1053–60.

[7] Modig J. Effectiveness of dextran 70 versus Ringer's acetate in traumatic shock and adult respiratory distress syndrome. Crit Care Med 1986;14(5):454–7.

[8] Human albumin administration in critically ill patients: systematic review of randomised controlled trials. BMJ 1998;317(7153):235–40.

[9] Lum LC, et al. Risk factors for hemorrhage in severe dengue infections. J Pediatr 2002;140(5):629–31.

[10] Thomas L, et al. Prospective observational study of low thresholds for platelet transfusion in adult dengue patients. Transfusion 2009;49(7):1400–11.

[11] Khan Assir MZ, et al. Effectiveness of platelet transfusion in dengue fever: a randomized controlled trial. Transfus Med Hemother 2013;40(5):362–8.

[12] Hariyanto H, et al. Management of severe dengue hemorrhagic fever and bleeding complications in a primigravida patient: a case report. J Med Case Rep 2016;10(1). 357-357.

[13] Leontiadis GI, Howden CW. The role of proton pump inhibitors in the management of upper gastrointestinal bleeding. Gastroenterol Clin North Am 2009;38(2):199–213.

[14] Thomas SJ, Endy TP. Critical issues in dengue vaccine development. Curr Opin Infect Dis 2011;24(5):442–50.

[15] Prompetchara E, et al. Dengue vaccine: global development update. Asian Pac J Allergy Immunol 2019.

[16] Swaminathan S, Khanna N. Dengue vaccine development: global and Indian scenarios. Int J Infect Dis 2019.

[17] Administration FD. First FDA-approved vaccine for the prevention of dengue disease in endemic regions. 2019. Available from: https://www.fda.gov/news-events/press-announcements/first-fda-approved-vaccine-prevention-dengue-disease-endemic-regions.

[18] Capeding MR, et al. Clinical efficacy and safety of a novel tetravalent dengue vaccine in healthy children in Asia: a phase 3, randomised, observer-masked, placebo-controlled trial. Lancet 2014;384(9951):1358–65.

[19] Villar L, et al. Efficacy of a tetravalent dengue vaccine in children in Latin America. N Engl J Med 2015;372(2):113–23.

[20] SANOFI. Sanofi updates information on dengue vaccine. 2017. Available from: https://www.sanofi.com/en/media-room/press-releases/2017/2017-11-29-17-36-30.

[21] Organization, W.H. Updated Questions and Answers related to information presented in the Sanofi Pasteur press release on 30 November 2017 with regards to the dengue vaccine Dengvaxia®. Immunization, Vaccines and Biologicals 2017 30th November 2017]; Available from:: https://www.who.int/immunization/diseases/dengue/q_and_a_dengue_vaccine_dengvaxia/en/.

[22] Soucheray, S. WHO advisers halt Dengvaxia, for now. 2018; Available from:: http://www.cidrap.umn.edu/news-perspective/2018/04/who-advisers-halt-dengvaxia-now.

[23] Moodie Z, et al. Neutralizing antibody correlates analysis of tetravalent dengue vaccine efficacy trials in Asia and Latin America. J Infect Dis 2018;217(5):742–53.

[24] Fatima K, Syed NI. Dengvaxia controversy: impact on vaccine hesitancy. J Global Health 2018;8(2). 010312-010312.

[25] Larson HJ, Hartigan-Go K, de Figueiredo A. Vaccine confidence plummets in the Philippines following dengue vaccine scare: why it matters to pandemic preparedness. Hum Vaccin Immunother 2019;15(3):625–7.

# Author Index

'*Note*: Page numbers followed by "f" indicate figures, "t" indicates tables.'

# Subject Index

'*Note*: Page numbers followed by "f" indicate figures, "t" indicates tables.'

## E



## F



## G

FSC
www.fsc.org
MIX
Papier | Fördert gute Waldnutzung
FSC® C083411

Kolibri 360 GmbH
Ferdinand-Jühlke-Straße 7
99095 Erfurt, Deutschland
produktsicherheit@kolibri360.de